The Extended Preferred Ordering Theorem for Radar Tracking Using the Extended Kalman Filter (Revised Edition)

The
Extended Preferred Ordering Theorem for Radar Tracking Using the Extended Kalman Filter

(Revised Edition)

Unbiased and Consistent Trajectory Estimation

Donald Myron Leskiw

Ming Lei Press

The Extended Preferred Ordering Theorem for Radar Tracking Using the Extended Kalman Filter (Revised Edition)

Unbiased and Consistent Trajectory Estimation

Designed and Published by Ming Lei Press
Front cover art by Lanjing Zhou

ISBN 9781092954518

0987654

Contents

Acknowledgements

This book is a product of my recent PhD dissertation. That research began, however, over forty years ago in New York City at the Riverside Research Institute (formerly the Electronics Research Laboratories of Columbia University). There, my boss Charles R. Pederson led me away from electromagnetics and radar cross-section modeling into the art and science of applied optimal estimation. And as I was a beginning Ph.D. student in mathematics at the time, Jurgen Moser of the Courant Institute showed me the interplay among the geometric, algebraic, and topological points-of-view of mathematics. Later, the Air Force's Richard R. Paul (soon to be General Paul) provided me with the most challenging tracking problems in ballistic missile defense and space warfare, and for several years I worked with Carl Kukkonen and David W. Curkendall of the Jet Propulsion Laboratory, and Geoffrey C. Fox and Thomas D. Gottschalk of Caltech on solutions – they provided the requisite techniques and high-performance technologies for their simulation, emulation, and demonstration. I also wish to acknowledge two colleagues whose reviews and criticisms of my work in radar tracking have been most useful over the years: Fred Daum of Raytheon and Ted Rice of Lockheed-Martin. But most of all, deep gratitude is given here to Kenneth S. Miller, formerly Professor of Mathematics at Columbia University and New York University. Both at Riverside and over the years hence, he taught me how to apply mathematics (also the importance of having a good notation). Together we wrote a book and several research papers on radar tracking. To Ken, advisor, mentor, and friend, this book is being offered now.
DML (Syracuse, NY)

1

Introduction

A certain problem in nonlinear estimation exists in radar tracking. Usually, the detections provide noisy measurements of the distance and direction from the radar to a target[1] while its position and motion are to be determined in Euclidean space (rectangular coordinates), and an estimator such as the Kalman filter is used. But most practitioners

> "... simply transform the polar coordinates to cartesian coordinates directly which modifies the noise process slightly. In this manner, the bias errors which build in the funny curvilinear coordinates is avoided and the filter gains are only suboptimal by trivial amounts due to the transformation ..."
>
> (anonymous IEEE reviewer of [1]).

In which case most radar tracks are biased and have inconsistent covariance matrices!

Of course, over the years many techniques have been proposed for mitigating those spurious estimation errors. Unfortunately, here we show that the leading ones can make them worse or require more detections to be effective. Fortunately, a simple method exists whereby the extended Kalman filter (EKF), which is a recursively linearized estimator, can provide an unbiased estimate that is less noisy, and with a consistent covariance

---

[1] In radar tracking parlance, it is customary to call an "object-of-interest" a target, regardless of whether it is a friend, foe, or neutral.

matrix. In the aforecited work by this author it was shown that an EKF's spurious linearization errors can be significantly reduced by using the measurement components of a detection in a certain order: angle first and range last. Such has since been dubbed the Preferred Ordering Theorem (POT) of radar tracking. It is used in real time tracking applications for air and missile defense [2], also non-realtime settings such as the Lincoln Orbit Determination (LODE) program of MIT's Lincoln Laboratory [3].

The objective of this monograph is to present a new version of the POT, the *Extended-Preferred Ordering Theorem (EPOT)*. Not only is the EPOT more efficient than the POT, using it the scalar measurement components of a detection can now be used in any order to update an EKF's track with virtually the same results. And under the EPOT the EKF becomes a stable estimator.

1.1 Background on Radar Tracking

The term "radar" was coined in 1940 by Lieutenant Commanders S. M. Tucker and F. R. Furth as an acronym for *radio detection and ranging* [4]. Most radars operate by radiating electromagnetic fields in certain directions and listening for their reflections from objects happen to be there. And if a suitable return is detected, they determine the reflector's distance from the radar as $r = \Delta t \times c/2$, where Δt is the travel time to the reflector and back, with c the speed of light. And an *angle* to the reflector is determined by receiving via a directional antenna, or beamforming.

Figure 1-1 provides a top-level block diagram of such a radar. Of course, the environment is usually unknown: a given return might be from one or more objects, clutter, or caused by interference and noise. The radar transmit certain waveforms and processes the returns to detect their presence. And tracking is how the radar determines which detections are associated with what objects, and which of those to report.

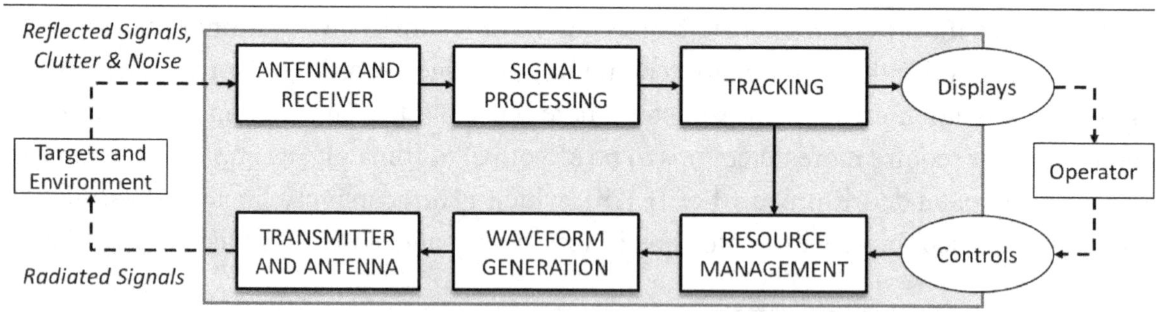

Figure 1-1 Radar system block diagram

Figure 1-2 illustrates the radar tracking process as a loop of basic four functions. *Track Prediction* propagates the tracks to the times of their next expected detections. *Correlation and Association* decides which new detections and what predicted tracks belong to the same object(s). *Track Update* corrects the predictions using their new associated detections. And *Track Management* decides whether an unassociated detection should be kept for initializing a new track, what model to use for predicting a track's next detection, whether a track should be reported, etc. Such a loop is also called a *filter*: the association decisions are made sequentially while the tracks are predicted and updated recursively, with their validity and classification reevaluated periodically.

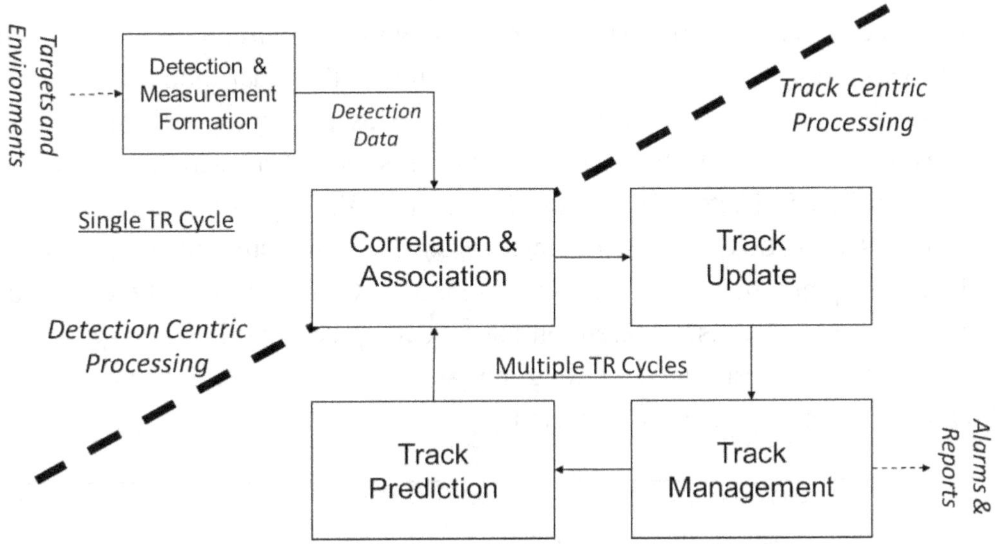

Figure 1-2 The basic tracking filter block diagram

The radar may also use its tracks to help manage its resources: for example, the accuracies of the measurements can be improved by predicting a target's position and then focusing the antenna and signal processing there. There are, however, two disparate classes of tracks. A *cooperative* target tells the radar its position and identity; its track is maintained to prevent its detections being confused with other targets. In contrast, the track of an *uncooperative* target is a hypothesis: a set of detections *believed* to be associated with the same object. Given the measurements in such a set, a model for the object's unknown motion is chosen for predicting its next detection, and if a successor detection is sufficiently close to that prediction, the track may be updated. In which case, the measurement errors and modeling approximations can lead to false tracks and missed targets. It is the latter case that mostly concerns us here.

But in either case, cooperative or not, a target may be difficult to detect, its become lost and need to be reacquired. If a target maneuvers unexpectedly, its model of motion might be changed. A stationary target might begin to move (the motion detection problem). And additional tracking challenges exist when multiple unresolved targets are present, also interference, jamming, and other radar tracking countermeasures.

Figure 1-3 illustrates a scenario for tracking an uncooperative target. There, a sequence of directed electromagnetic pulses called *beams* is depicted as cones – the vertices are at the radar, the widths are proportional to the directivity of the antenna. The notion of a *range gate* is also shown, the shaded regions of Beams 2 and 3: the entirety of Beam 1 is searched for new targets; just the range gates of the other beams are searched for the previously detected one. To initialize this track, let an *unassociated* detection be given in Beam 1. To *verify* it is not a false alarm caused by noise, the radar then transmits a cluster of beams covering an area where the target could be, assuming a maximum speed for its motion. Beam 2 is successful (the unsuccessful beams are not shown). The measurements from the first and second detections are used to estimate the object's position and velocity, and the object's third detection is predicted. Beam 3 is transmitted, with a range gate centered at that prediction, and an associated detection is found there, presumably *validating* those assumptions. Whereupon the object's position and velocity are updated. The acceleration is also estimated, leading the radar to classify the object as an incoming ballistic missile, a new target to be reported.

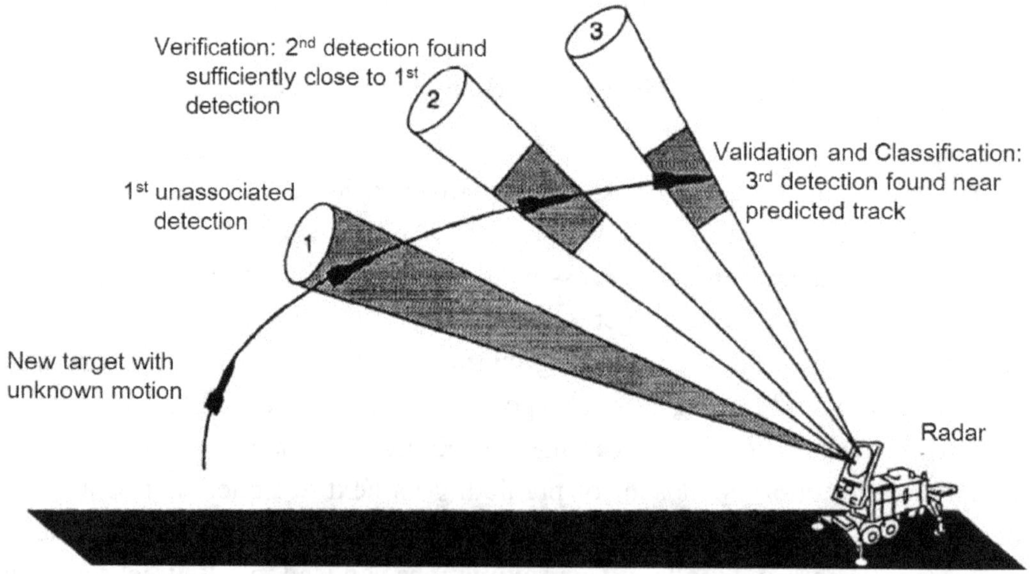

Figure 1-3 Initializing a radar track on an uncooperative target [5]

1.2 Radar Tracking in Euclidean Space

In this work we define the problem in Euclidean space, which is coordinateless, and then analyze its radar and rectangular coordinate representations. Accordingly, Figure 1-4 illustrates the construction of a Euclidean track for the above scenario [6]. There the measurements are given as coordinateless points marked by **x**'s – to be more explicit, however, we also label them as $\overline{P}_1$ $\overline{P}_2, \overline{P}_3$, corresponding to the beams. The estimates are marked by circles, with hollow ones, **o**, for the predictions and solid ones, •, for the updates. The first two estimates are $\overline{P}_1$ and $\hat{P}_2 = \overline{P}_2$. The Euclidean prediction for Beam 3 is denoted $\hat{P}_3'^-$, and obtained by extending the line segment from $\overline{P}_1$ to $\overline{P}_2$ by the distance from $\overline{P}_1$ to $\overline{P}_2$. Its update is at $\hat{P}_3'$, located on the line segment *between* $\hat{P}_3'^-$ and $\overline{P}_3$, five normalized units from $\hat{P}_3'^-$, or one from $\overline{P}_3$ (our use of primes is explained below – the 5/6 shall be derived in Chapter 2). Of course, to qualify as a *radar* track, the measurements must be given as *distances* (range) and *directions* (angles) from the radar, not coordinateless Euclidean points.

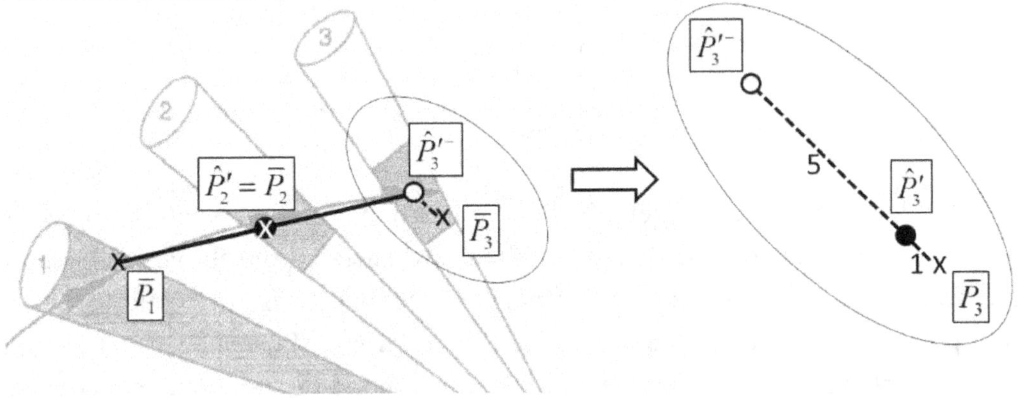

Figure 1-4 Construction of an initial Euclidean track

Figure 1-5 provides the constructions of the corresponding tracks in radar and rectangular coordinates. On the left-hand side the range and angle measurements are given as circles and radials – their intersections are at the same $\overline{P}_n$, of Figure 1-4, but now labeled as $(\overline{r}_n, \overline{a}_n)$, $n = 1,2,3$. And the first two estimates are the same as in Figure 1-4; the prediction for Beam 3 is different. Its *range* is constructed by reflecting the radial segment of length $\overline{r}_1 - \overline{r}_2$ about the circle of radius $\overline{r}_2$, that is, $\hat{r}_3^- = \overline{r}_2 + (\overline{r}_2 - \overline{r}_1)$; its *angle* is obtained by reflecting the radial at $\overline{a}_1$ about the radial at $\overline{a}_2$, that is , $\hat{a}_3^- = \overline{a}_2 + (\overline{a}_2 - \overline{a}_1)$. Their intersection is at $(\hat{r}_3^-, \hat{a}_3^-)$. The update, however, requires the trisection of an angle, so it is not constructed here (we determine it below, algebraically). The right-hand side of Figure 1-5 shows the corresponding rectangular coordinate case – it is basically the same as Figure 1-4, since rectangular coordinate space and Euclidean space are isomorphic. In the latter case in Figure 1-5, we have used geometric vectors denoted by $\boldsymbol{x}$'s, and the parallelogram rule for adding and subtracting them. Thus, the prediction for Beam 3 is labeled as $\hat{\boldsymbol{x}}_3'^- = \bar{\boldsymbol{x}}_2 + (\bar{\boldsymbol{x}}_2 - \bar{\boldsymbol{x}}_2)$. Its update would be $\hat{\boldsymbol{x}}_3' = \hat{\boldsymbol{x}}_3'^- + 5(\bar{\boldsymbol{x}}_3 - \hat{\boldsymbol{x}}_3'^-)/6$. Such are *free* vectors – they may be translated freely.

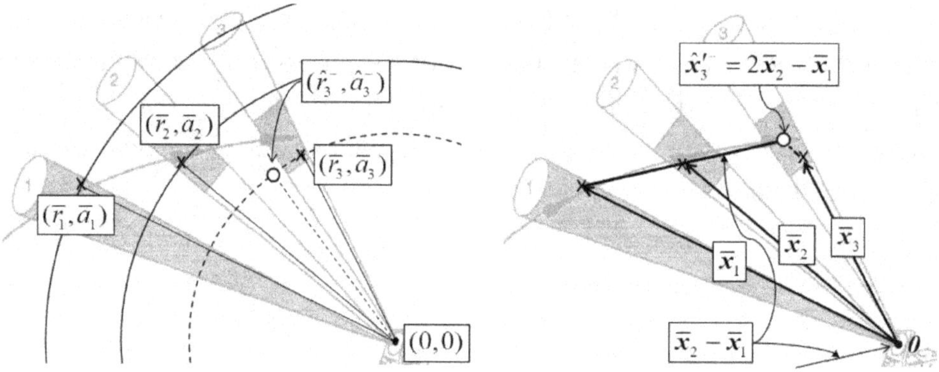

Figure 1-5 The principal predictions in Euclidean space

To determine these tracks algebraically, we use linear algebra vectors, to be denoted by $\mathbf{r}$'s and $\mathbf{x}$'s. Such are column vectors of the coordinates for the above Euclidean points and are *bound* to the origin of their respective vector spaces. In particular, given an arbitrary point in Euclidean space, P, let its true radar coordinates be (r, a), with $r > 0$ and $-\pi < a \le \pi$. And let them be related to the corresponding rectangular coordinates as

$$\begin{bmatrix} x \\ y \end{bmatrix} = \begin{bmatrix} r\cos a \\ r\sin a \end{bmatrix} \quad \text{and} \quad \begin{bmatrix} r \\ a \end{bmatrix} = \begin{bmatrix} (x^2+y^2)^{1/2} \\ \arctan(x, y) \end{bmatrix}. \tag{1}$$

Note that the four-quadrant inverse tangent function is being used here, $-\pi < a \le +\pi$, not the two quadrant one, $\phi \equiv \arctan(\sin\phi / \cos\phi)$, $-\pi/2 < \phi \le +\pi/2$.

Now let the measurement points in Figure 1-4, $\bar{P}_n$, $n = 1,2,3$, be given as $(\bar{r}_n, \bar{a}_n)$, with their converted values obtained via (1), namely, $(\bar{x}_n, \bar{y}_n)$. The initial radar and rectangular estimates are then $(\hat{r}_1, \hat{a}_1) = (\bar{r}_1, \bar{a}_1)$ and $(\hat{x}_1, \hat{y}_1) = (\bar{x}_1, \bar{y}_1)$, without the primes, both representing $\hat{P}_1 = \bar{P}_1$. And the predictions at $n = 2$ are $(\hat{r}_2^-, \hat{a}_2^-) = (\hat{r}_1, \hat{a}_1)$ and $(\hat{x}_2^-, \hat{y}_2^-) = (\hat{x}_1, \hat{y}_1)$, both representing $\hat{P}_2^- = \hat{P}_1$. Using the measurements at times t_1 and t_2

we then determine *midpoint estimates* at $t_m = (t_1 + t_2)/2$, namely,

$$\begin{bmatrix} \hat{r}_m \\ \hat{a}_m \end{bmatrix} = \frac{1}{2}\begin{bmatrix} \bar{r}_1 + \bar{r}_2 \\ \bar{a}_1 + \bar{a}_2 \end{bmatrix} \text{ and } \begin{bmatrix} \hat{x}'_m \\ \hat{y}'_m \end{bmatrix} = \frac{1}{2}\begin{bmatrix} \bar{x}_1 + \bar{x}_2 \\ \bar{y}_1 + \bar{y}_2 \end{bmatrix}. \tag{2}$$

The first expression in (2) is the midpoint along a *spiral* in Euclidean space, between $\bar{P}_1$ and $\bar{P}_2$, while the second expression determines the midpoint on the Euclidean *line* between $\bar{P}_1$ and $\bar{P}_2$. Next, propagate those midpoint estimates to t_2 as

$$\begin{bmatrix} \hat{r}_2 \\ \hat{a}_2 \end{bmatrix} = \begin{bmatrix} \hat{r}_m \\ \hat{a}_m \end{bmatrix} + \frac{t_2 - t_m}{t_2 - t_1}\begin{bmatrix} \bar{r}_2 - \bar{r}_1 \\ \bar{a}_2 - \bar{a}_1 \end{bmatrix} \text{ and } \begin{bmatrix} \hat{x}'_2 \\ \hat{y}'_2 \end{bmatrix} = \begin{bmatrix} \hat{x}'_m \\ \hat{y}'_m \end{bmatrix} + \frac{t_2 - t_m}{t_2 - t_1}\begin{bmatrix} \bar{x}_2 - \bar{x}_1 \\ \bar{y}_2 - \bar{y}_1 \end{bmatrix}. \tag{3}$$

After substituting (2) into (3) we have

$$\begin{bmatrix} \hat{r}_2 \\ \hat{a}_2 \end{bmatrix} = \begin{bmatrix} \bar{r}_2 \\ \bar{a}_2 \end{bmatrix} \text{ and } \begin{bmatrix} \hat{x}'_2 \\ \hat{y}'_2 \end{bmatrix} = \begin{bmatrix} \bar{x}_2 \\ \bar{y}_2 \end{bmatrix}. \tag{4}$$

For Beam 3, the predictions are then determined by propagating the estimates in (4) as

$$\begin{bmatrix} \hat{r}_3^- \\ \hat{a}_3^- \end{bmatrix} = \begin{bmatrix} \hat{r}_2 \\ \hat{a}_2 \end{bmatrix} + \frac{t_3 - t_2}{t_2 - t_1}\begin{bmatrix} \bar{r}_2 - \bar{r}_1 \\ \bar{a}_2 - \bar{a}_1 \end{bmatrix} \text{ and } \begin{bmatrix} \hat{x}_3'^- \\ \hat{y}_3'^- \end{bmatrix} = \begin{bmatrix} \hat{x}_2 \\ \hat{y}_2 \end{bmatrix} + \frac{t_3 - t_2}{t_2 - t_1}\begin{bmatrix} \bar{x}_2 - \bar{x}_1 \\ \bar{y}_2 - \bar{y}_1 \end{bmatrix}. \tag{5}$$

And when $t_3 - t_2 = t_2 - t_1$,

$$\begin{bmatrix} \hat{r}_3^- \\ \hat{a}_3^- \end{bmatrix} = \begin{bmatrix} \hat{r}_2 \\ \hat{a}_2 \end{bmatrix} + \begin{bmatrix} \bar{r}_2 - \bar{r}_1 \\ \bar{a}_2 - \bar{a}_1 \end{bmatrix} \text{ and } \begin{bmatrix} \hat{x}_3'^- \\ \hat{y}_3'^- \end{bmatrix} = \begin{bmatrix} \hat{x}_2 \\ \hat{y}_2 \end{bmatrix} + \begin{bmatrix} \bar{x}_2 - \bar{x}_1 \\ \bar{y}_2 - \bar{y}_1 \end{bmatrix}. \tag{6}$$

Note that in the radar coordinate case, the sums and differences in angle must be *corrected for branch cut*: they are determined such that their angles are in the interval $-\pi < a \le +\pi$ with $r > 0$. Substituting the expressions in (4) into those of (6), the coordinates of the constructions in Figure 1-5 at $\hat{P}_3^-$ and $\hat{P}_3'^-$ are now found to be

$$\begin{bmatrix} \hat{r}_3^- \\ \hat{a}_3^- \end{bmatrix} = 2\begin{bmatrix} \bar{r}_2 \\ \bar{a}_2 \end{bmatrix} - \begin{bmatrix} \bar{r}_1 \\ \bar{a}_1 \end{bmatrix} \text{ and } \begin{bmatrix} \hat{x}_3'^- \\ \hat{y}_3'^- \end{bmatrix} = 2\begin{bmatrix} \bar{x}_2 \\ \bar{y}_2 \end{bmatrix} - \begin{bmatrix} \bar{x}_1 \\ \bar{y}_1 \end{bmatrix}. \tag{7}$$

Again, the radar coordinate case determines a point on a *spiral* relative to the radar, hence $\hat{P}_3^- \ne \hat{P}_3'^-$ (unless $\bar{a}_3 = \bar{a}_2$), while the rectangular case is alone a line. And finally, given $(\bar{r}_3, \bar{a}_3)$ and $(\bar{x}_3, \bar{y}_3)$, the updates are

$$\begin{bmatrix}\hat{r}_3\\ \hat{a}_3\end{bmatrix}=\begin{bmatrix}\hat{r}_3^-\\ \hat{a}_3^-\end{bmatrix}+\frac{5}{6}\begin{bmatrix}\overline{r}_3-\hat{r}_3^-\\ \overline{a}_3-\hat{a}_3^-\end{bmatrix} \text{ and } \begin{bmatrix}\hat{x}_3'\\ \hat{y}_3'\end{bmatrix}=\begin{bmatrix}\hat{x}_3'^-\\ \hat{y}_3'^-\end{bmatrix}+\frac{5}{6}\begin{bmatrix}\overline{x}_3-\hat{x}_3'\\ \overline{y}_3-\hat{y}_3'\end{bmatrix} \quad (8)$$

The radar coordinate case, $(\hat{r}_3,\hat{a}_3)$, represents $\hat{P}_3$, between $\hat{P}_3^-$ and $\overline{P}_3$ along a *spiral*. The rectangular coordinate case, $(\hat{x}_3',\hat{y}_3')$, represents the Euclidean update, $\hat{P}_3'$, on the *line* between $\hat{P}_3'^-$ and $\overline{P}_3$. (In Chapter 2 we will discuss other models, in addition to the spirals and lines being used here, and the so-called Kalman *system noise* is introduced.)

1.3 Illustration of the Spurious Estimation Errors

Now consider a special case where the target is stationary, and let a sequence of unbiased, mutually independent radar measurements on P be given, $(\overline{r}_n,\overline{a}_n)$, $n=1,2,\cdots,N$, whose errors are Gaussian. By definition, the converted values are $\overline{x}_n=\overline{r}_n\cos\overline{a}_n$ and $\overline{y}_n=\overline{r}_n\sin\overline{a}_n$. And then determine the estimates as

$$\begin{bmatrix}\hat{r}_n\\ \hat{a}_n\end{bmatrix}=\frac{1}{n}\sum_{i=1}^{n}\begin{bmatrix}\overline{r}_i\\ \overline{a}_i\end{bmatrix} \text{ and } \begin{bmatrix}\hat{x}_n'\\ \hat{y}_n'\end{bmatrix}=\frac{1}{n}\sum_{i=1}^{n}\begin{bmatrix}\overline{x}_i\\ \overline{y}_i\end{bmatrix}. \quad (9)$$

Such are estimation sequences that presumably get better as n increases. Figure 1-6 illustrates this case for $n=1,2,\cdots,N$, with $N=10{,}000$, where $(r,a)=(x,y)=(1,0)$.

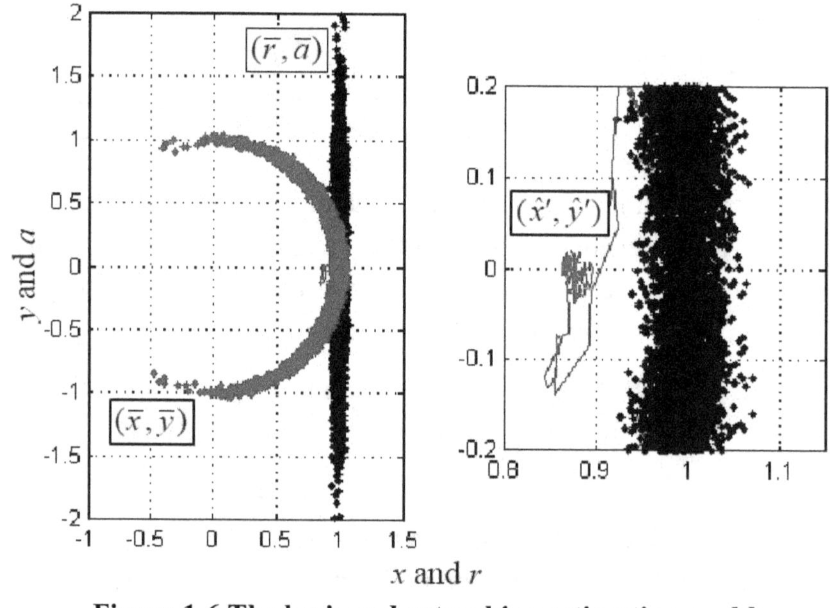

Figure 1-6 The basic radar tracking estimation problem

In the above figure, the data and estimates in radar coordinates are overlaid upon the same coordinate axes – the right-hand side is an enlargement. To make their comparison transparent, we have chosen the true coordinates to be the same: the set of Gaussian radar measurements has a symmetrical, elliptical shape centered at $(1,0)$, with major axis parallel to the vertical axis; the set of converted measurements follows a semi-circle centered at $(0,0)$. Since $r=1$, the angle a plays the role of *cross-range* – as does y when $a=0$. And to make their disparities obvious, the range and angle measurement accuracies have standard deviations $\sigma_r = .02$ meters and $\sigma_a = \pi/6$ radians, with covariance $\sigma_{ra} = 0$.

In particular, the estimates in rectangular coordinates are clearly seen because they are *biased*; the estimates in radar coordinates are obscured by the data. Not obvious is that the $(\hat{x}', \hat{y}')$'s are noisier than the $(\hat{r}, \hat{a})$'s. This aspect of the problem is illustrated below in Figure 1-7. There the $(\overline{r}_n, \overline{a}_n)$ and $(\overline{x}_n, \overline{y}_n)$, $n = 1,2,\cdots,N$, have been reindexed as $(\overline{r}_{l,m}, \overline{a}_{l,m})$ and $(\overline{x}_{l,m}, \overline{y}_{l,m})$ with $l = 1,2,\cdots,L$ and $m = 1,2,\cdots,M$, where $L = 200$ and $M = 50$. Each curve is an independent Monte Carlo trial of length L, determined as

$$\begin{bmatrix} \hat{r}_{l,m} \\ \hat{a}_{l,m} \end{bmatrix} = \frac{1}{l}\sum_{k=1}^{l} \begin{bmatrix} \overline{r}_{k,m} \\ \overline{a}_{k,m} \end{bmatrix} \quad \text{and} \quad \begin{bmatrix} \hat{x}'_{l,m} \\ \hat{y}'_{l,m} \end{bmatrix} = \frac{1}{l}\sum_{k=1}^{l} \begin{bmatrix} \overline{x}_{k,m} \\ \overline{y}_{k,m} \end{bmatrix}, \tag{10}$$

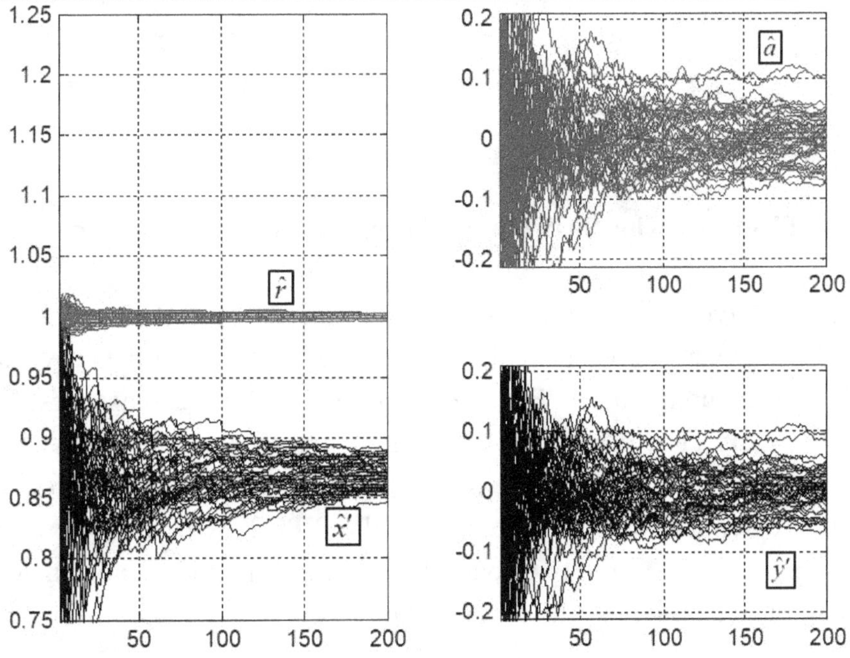

Figure 1-7 The basic estimation paths

Of course, the final estimates in (9) are the same as the final estimates in (10),

$$\begin{bmatrix}\hat{r}_N\\ \hat{a}_N\end{bmatrix}=\frac{1}{M}\sum_{i=1}^{m}\begin{bmatrix}\bar{r}_{L,m}\\ \bar{a}_{L,m}\end{bmatrix}\quad\text{and}\quad\begin{bmatrix}\hat{x}'_N\\ \hat{y}'_N\end{bmatrix}=\frac{1}{M}\sum_{i=1}^{m}\begin{bmatrix}\bar{x}_{L,m}\\ \bar{y}_{L,m}\end{bmatrix},\tag{11}$$

when $N=LM$. Obviously, any permutation of the indices does not affect the final outcomes at N.

One attribute they share is that they are always *between* their constituent summands. Figure 1-4 illustrated this "betweenness" property: the Euclidean update is between $\hat{P}_3'^-$ and $\bar{P}_3$. And in (8), the updates $(\hat{r}_3,\hat{a}_3)$ and $(\hat{x}_3'^-,\hat{y}_3'^-)$ are between the predictions and successor measurements. And Figure 1-8 demonstrates it for the special case where the measurement in Figure 1-6 have been reindexed such that $\bar{a}_1=\bar{a}_{\min}$ and $\bar{a}_2=\bar{a}_{\max}$.

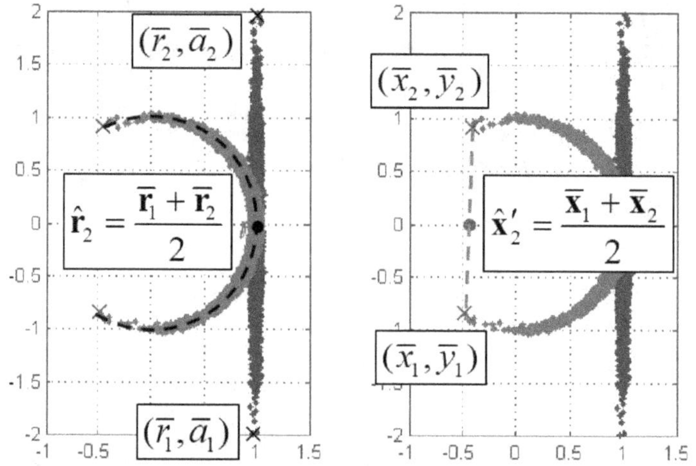

Figure 1-8 The "betweenness" property of the linear estimators

Note that the converted measurements are not biased *per se*. (They do not need to be "debiased"!) The converted case provides an unbiased estimate of the barycenter (center-of-mass) of the measurements in Euclidean space, which is a biased estimator of the target's position there. The direct case provides a biased estimate of that barycenter; it instead provides the "barycenter" of the measurement points in radar coordinate space. In the sequel, the extended Kalman filter with the EPOT will exploit these disparate properties of the estimator in the two spaces: it shall use the radar coordinates directly to determine an unbiased update of the predicted track in rectangular coordinates.

1.4 Background on the Problem to Be Discussed

The Kalman filter is basically a recursive linear estimator with a quadratic cost function, with the unknowns modeled as a linear, mean-squared continuous random process [7]. (The popular alpha-beta radar tracking filter is a steady-state Kalman filter [8].) The estimation equations are those of weighted least-squares: they provide the Best Linear Unbiased Estimate (BLUE) when the true inverse covariance matrices are used as the weights. And when all the models are linear and all the errors have Gaussian distributions with zero means, the BLUE provides the Gauss-Markov estimate, which is optimal in the *unbiased* and *minimum variance* sense [9].

Most radar tracking problems are *nonlinear* and *non-gaussian*. The measurements are given in radar coordinates, with *range* always positive and the *angles* limited to their principal values. And the modeling errors of a target's unknown motion tend to be systematic. The Gaussian assumption is weakly motivated by the Central Limit Theorem; it is strongly motivated by the tractability it gives to an analysis [10].

Early on, Schmidt (1962) suggested that the models be linearized recursively, which became known as the *extended Kalman filter* (EKF) [11], and Jazwinski (1970) proposed that the linearization errors be modeled using the so-called *system noise* [12] – for others see pp. 5-9 of [13]. But the EKF's linearization errors are also systematic: they tend to accumulate and can lead a track to diverge from its object. A "second-order" might be used, or an "iterated" EKF, but they require more computations, and still have systematic errors (albeit, higher-order ones) [14].

There are *bone fide* nonlinear estimators such as a *particle filter*, which can provide "statistically linearized" estimates that are optimal in the limit [15]. But they are also "computationally intensive" [16]. An alternative is the Unscented Kalman Filter (UKF), which models certain parameters of the measurement probability distribution, and transforms those within a Kalman filtering framework [17]. And others attempt to "debias" the transformed measurements before they are used. For example, the Debiased Consistent Converted Measurements (DCCM) method subtracts the expected value of the estimation errors from the transformed measurements [18]. The Unbiased Consistent Converted Measurements (UCCM) method dilates the transformed measurements first [19]. But they also require knowledge of the underlying probability distribution of the measurements, which is generally unknown in practice. And so most radar tracks are *biased* [20], and their covariance matrices determined by the Kalman filter are *inconsistent* with the estimation errors [21].

Fortunately, a certain noncommutativity exists when the components of a radar measurement are used recursively to update an EKF's track in rectangular coordinates. There is a preferred order: azimuth first and range last [3]. Here that method is called the Preferred Ordering Theorem (POT). Unfortunately, the POT is counterintuitive to a common belief that the most accurate measurement should be used first, so as to obtain a better linearization (see p. 166, [22]) – such implies the worst order, since range measurements are usually much more accurate than the angle ones (in cross-range).

The original motivation of the POT was a Best Estimate of Trajectory (BET) problem for an advanced maneuvering reentry vehicle. Since then, it has been used in real time tracking applications [2], also in non-realtime settings like the Lincoln Orbit Determination (LODE) program of MIT Lincoln Laboratory [3]. And others have successfully combined it with the DCCM/ UCCM [23, 24], and with the UKF [25].

1.5 Outline of the Sequel

This monograph is organized into eight Chapters, plus an Appendix. And to expedite the presentation, the discussion is mostly restricted to the Euclidean plane.

Chapter 1 provides this introduction. Chapter 2 presents the basic estimation equations. The basic notation is specified, and four canonical radar tracking cases are defined. In Chapter 3 the detailed equations of those tracking filters are presented and the spurious estimation errors they cause are demonstrated. Chapter 4 presents an error analysis of the basic estimation errors. The situation where the popular "debiasing" methods make the biases worse is identified and a remedy for that is given. Chapter 5 provides a new analysis of the EKF errors. The EKF with the POT is illustrated and seen to be less effective at short ranges. The *Basic*-EPOT (B-EPOT) is then defined and demonstrated to address the latter problem. After that, Chapter 6 uses the methods from the previous Chapters in a more stressing tracking problem, taken from the DCCM/UCCM literature, and all the trackers are seen to have problems, including the B-EPOT. Chapter 7 analyzes those errors, and a certain *position-velocity consistency constraint* is derived. Using it, the EPOT is then complete. And finally, Chapter 8 provides some concluding remarks.

1.6 Chapter 1 References

[1] K. S. Miller and D. M. Leskiw, "Nonlinear Estimation with Radar Observations," in Transitions on Aerospace and Electronic Systems, IEEE AES-2, pp. 192-200 (1982).

[2] L. Nelsen T. Carlson, "Techniques for Robust Tracking in Airborne Radar," 89CH2685-6/89/0000-0080 (1989).

[3] D. Leskiw and C. Pederson, Metric Accuracy Analysis, Riverside Research Institute Technical Report ESD-TR-79-332 (1978).

[4] R. Buderi: The Invention That Changed the World, Simon & Schuster (1996).

[5] Patriot Missile Defense: Software Problem Led to System Failure at Dhahran, Saudi Arabia, United States General Accounting Office, IMTEC-92-26 (Publicly Released February 27, 1992).

[6] N. D. Kazarinoff: Ruler and the Round: Classic Problems in Geometric Constructions, Dover Publications (2011).

[7] T. Kailath, Linear Estimation, Prentice Hall (2000).

[8] E. Brookner, Tracking and Kalman Filtering Made Easy, Wiley (1998).

[9] K. S. Miller and D. M. Leskiw, An Introduction to Kalman Filtering With Applications, Krieger (1987).

[10] S. Rowels, and Z, Ghahramani, "A Unifying Review of Linear Gaussian Models," in Neural Computation II, pp. 305-345, MIT Press (1999).

[11] G. Smith, S. Schmidt, and L. McGee, "Application of Statistical Filtering to the Optimal Estimation of Position and Velocity On board a Circumlunar Vehicle," NASA Ames Research Center Report, NASA TND-1205 (1962).

[12] A. Jazwinski, "Adaptive Filtering," Automatica, Vol. 5, pp. 475-785 (1969).

[13] G. J. Bierman, Factorization Methods for Discrete Sequential Estimation, Academic Press (1977).

[14] A. Gelb (Ed), Applied Optimal Estimation, MIT Press (1974).

[15] B. Ristic, S. Arulampalam, and N. Gordon, Beyond the Kalman Filter; Particle Filters for Tracking Applications, Artech (1995).

[16] F. Daum, Co., "Nonlinear Filters: Beyond the Kalman Filter," in IEEE Aerospace and Electronic Systems Magazine, Vol-20 Issue 8, pp. 57-69 (August 2005).

[17] S. Julier and J. Uhlmann, "A New Extension of the Kalman Filter to Nonlinear Systems," in Proceedings of AeroSense: Symposium Aerospace/Defense Sensing, Simulation and Control, Vol. 3373, pp. 183-193 (1997).

[18] D. Lerro and T. Bar-Shalom, "Tracking with Debiased Consistent Converted Measurements versus EKF," in Transactions on Aerospace and Electronic Systems, IEEE AES-29, pp. 1015-1022 (1993).

[19] L. Mo, X. Song, Y. Zhou, and Z. Sun, "Alternative Unbiased Consistent Converted Measurements for Target Tracking," in Proceedings of Acquisition, Tracking, and Pointing XI, M. K. Masten and L. A. Stockum (eds.), SPIE Vol. 3086, p. 308-310 (1997).

[20] M. Miller and O. Drummond, "Coordinate Transformation Bias in Target Tracking," in Signal and Data Processing of Small Targets, SPIE Vol. 3809, pp. 409-424 (1999).

[21] O. Drummond, A. Perrella, S. Waugh, "On Target Track Covariance Consistency," in Signal and Data Processing of Small Targets, SPIE Vol. 6236 (2006).

[22] S. S. Blackman and R. Popoli, Design and Analysis of Modern Tracking Systems, Artech (1999).

[23] S. Park J. Lee, "An Efficient Filtering Algorithm for Improved Radar Tracking," in IEEE Conference on Control Applications, pp. 1078-1083 (1996).

[24] S. Park J. Lee, "Improved Kalman Filter Design for Three-Dimensional Radar Tracking," in Transactions on Aerospace and Electronic Systems, IEEE AES-37 pp. 727-723 (2001).

[25] Z. Daun, X. Li, C. Han, and C. Hong, "Sequential Unscented Kalman Filter for Radar Target Tracking with Range Rate Data," 0-7803-9289-8/05 (2005).

2

Mathematical Preliminaries

The purpose of this Chapter is to present the basic radar tracking equations and define the notation that shall be used in the sequel. Whereas the focus of the analysis will be upon the spurious linear estimation errors that are caused by the nonlinear transformation between radar and rectangular coordinates, we begin with "coordinateless" Euclidean space. That is then followed by its radar and rectangular coordinate representations, and the transformations between them. The basic tracking equations are then given. The chapter concludes with a summary of the various estimators to be used in the sequel, and the notation is summarized in the conclusion.

2.1 The Principal Spaces

Recall that Euclidean space, here denoted by $\mathbb{E}$, is homogeneous and isotropic, with no point or direction inherently distinguished. Its points and sets of points may be chosen freely. And there the common notions of distance and angle are invoked axiomatically, along with the translation, rotation, reflection isometries [1-5].

Accordingly, the radar is said to be a certain fixed point, $O \in \mathbb{E}$. And the object to be tracked is another point, P, which is arbitrary except for $P \neq O$. Denote the directed line segment in $\mathbb{E}$ from O to P by OP. Write the distance between O and P as $\|OP\|$ (a unit length is tacitly chosen in $\mathbb{E}$); and represent the orientation and positive sense of OP by an abstract unit vector, $\boldsymbol{e}_{OP}$ (such is defined more formally at the end of this section).

Now choose a fixed rectangular coordinate frame in $\mathbb{E}$, denoted by $[O;(\boldsymbol{e}_x,\boldsymbol{e}_y)]$, where O specifies the origin, and $(\boldsymbol{e}_x,\boldsymbol{e}_y)$ is an ordered pair of orthonormal (abstract) unit vectors that represent the orientation and positive sense of the coordinate axes. Under $[O;(\boldsymbol{e}_x,\boldsymbol{e}_y)]$ the *rectangular coordinates* of the radar and object are respectively $(0,0)$ and (x,y). And, given (x,y), with $(x,y)\neq(0,0)$, the *radar coordinates* of the object, *range* and *azimuth*, are defined to be

$$r=(x^2+y^2)^{1/2} \quad \text{and} \quad a\equiv\arctan(y,x) \tag{12}$$

(the four-quadrant inverse tangent function is being used here, $-\pi<a\leq+\pi$, not the two quadrant one, $\phi\equiv\arctan(\sin\phi/\cos\phi)$, $-\pi/2<\phi\leq+\pi/2$). In which case, $r=\|OP\|$ and $\boldsymbol{e}_{OP}=\boldsymbol{e}_x\cos a+\boldsymbol{e}_y\sin a$. And given (r,a) as the radar coordinates of $P\in\mathbb{E}$, the corresponding rectangular coordinates are

$$x=r\cos a \quad \text{and} \quad y=r\sin a\,. \tag{13}$$

Finally, r and a in (13) may have any values, and so the phrase "the *principal values* of (r,a)" shall be used to indicate $r>0$ and $-\pi<a\leq+\pi$.

Now for arbitrary radar and rectangular coordinates, say (r,a) and (x,y), not necessarily related by (12) and (13), let them respectively correspond to the points R and X in $\mathbb{E}$. Either $X=R$ or $X\neq R$. If $X=R$, then usually $(r,a)\neq(x,y)$; and if $(r,a)=(x,y)$, then usually $X\neq R$. Of course, when (x,y) and (r,a) are related by (12) and (13), then $X=R$. But a special case exists: $X=R\Leftrightarrow(r,a)=(x,y)$ if and only if $x=r\geq0$ and $y=a=0$. This special case shall be used in the sequel to illustrate the effects of the nonlinear coordinate transformations upon the estimators.

Column vectors of radar and rectangular coordinates shall also be used to represent points in $\mathbb{E}$, namely,

$$\boldsymbol{\rho}=\begin{bmatrix} r \\ a \end{bmatrix} \quad \text{and} \quad \boldsymbol{\xi}=\begin{bmatrix} x \\ y \end{bmatrix}. \tag{14}$$

Such are members of distinct vector spaces where the standard basis is used, say $\boldsymbol{\rho}\in\boldsymbol{R}$ and $\boldsymbol{\xi}\in\boldsymbol{X}$. Their transposes will be written as $\boldsymbol{\rho}^T=(r,\ a)$ and $\boldsymbol{\xi}^T=(x,\ y)$, with a space after the comma to distinguish them from their corresponding coordinate points. For convenience, when the components of $\boldsymbol{\rho}$ and $\boldsymbol{\xi}$ in (14) are related by (12) and (13), those vectors shall be said to be functions of one another, $\boldsymbol{\rho} = \mathbf{h}(\boldsymbol{\xi})$ and $\boldsymbol{\xi} = \mathbf{h}^{-1}(\boldsymbol{\rho})$. Of course, $\mathbf{h}^{-1}$ is not generally the inverse of $\mathbf{h}$ per se; only if the radar coordinates are restricted to their principal values are $\mathbf{h}$ and $\mathbf{h}^{-1}$ inverses of one another.

Finally, the (abstract) unit vectors $\boldsymbol{e}_{OP}$, $\boldsymbol{e}_x$, $\boldsymbol{e}_y$, used above to represent the orientation and positive sense of directed line segments in $\mathbb{E}$, are not members of $\boldsymbol{X}$ or $\boldsymbol{R}$. Rather, they are members of an affine space that is isomorphic to $\mathbb{E}$, written ($\mathbb{E}$; $\boldsymbol{X}$) [5], simply denoted by $\mathbf{A}$. Objects in $\mathbf{A}$ shall always be denoted using lowercase bold-italic symbols. And the symbol "$\boldsymbol{e}$" shall always denote unit vectors there. The null vector in $\mathbf{A}$ and the column vector of zeros in $\boldsymbol{X}$ and $\boldsymbol{R}$ shall all be denoted by $\mathbf{0}$. Note that $\mathbb{E}$, $\boldsymbol{X}$, $\mathbf{A}$ are isomorphic to one another, while $\boldsymbol{R}$ is not isomorphic to any of them.

For example, given $\boldsymbol{\xi}$ and $\boldsymbol{\rho}$ in (14), the corresponding objects in $\mathbf{A}$ are written $\boldsymbol{\xi}$ and $\boldsymbol{\rho}$. When (r,a) and (x,y) are related by (12) and (13), then

$$\boldsymbol{\xi} \equiv x\boldsymbol{e}_x + y\boldsymbol{e}_y = (x,\ y)\begin{bmatrix} \boldsymbol{e}_x \\ \boldsymbol{e}_y \end{bmatrix} \text{ and } \boldsymbol{\rho} \equiv r\boldsymbol{e}_r(a) + 0\boldsymbol{e}_a(a) = (r,\ 0)\begin{bmatrix} \boldsymbol{e}_r(a) \\ \boldsymbol{e}_a(a) \end{bmatrix}, \tag{15}$$

with

$$\begin{bmatrix} \boldsymbol{e}_r(a) \\ \boldsymbol{e}_a(a) \end{bmatrix} = \begin{bmatrix} \cos a & +\sin a \\ -\sin a & \cos a \end{bmatrix}\begin{bmatrix} \boldsymbol{e}_x \\ \boldsymbol{e}_y \end{bmatrix} \tag{16}$$

(the arguments of $\boldsymbol{e}_r(a)$ and $\boldsymbol{e}_a(a)$ will be dropped when the context allows).

Denote the matrix in (16) by $\mathbf{O}(a)$. It represents a (positive) rotation in $\mathbb{E}$ about the origin, O, by the azimuth angle, a. Under $\mathbf{O}(a)$ the $[O;(\boldsymbol{e}_x,\boldsymbol{e}_y)]$ frame is rotated onto another rectangular coordinate frame, $[O;(\boldsymbol{e}_r,\boldsymbol{e}_a)]$. Note that $\mathbf{O}(a)$ is orthonormal, $\mathbf{O}^{-1}(a) = \mathbf{O}^T(a)$, and that $\mathbf{O}^{-1}(a) = \mathbf{O}(-a)$.

Given (x,y) and (r,a), respectively the rectangular and radar coordinates of X and R in $\mathbb{E}$, when $X = R$ then $\boldsymbol{\xi} = \boldsymbol{\rho}$, while $\boldsymbol{\xi} = \boldsymbol{\rho}$ need not be true. Indeed,

$$\boldsymbol{\xi} = (x,\ y)\begin{bmatrix} \boldsymbol{e}_x \\ \boldsymbol{e}_y \end{bmatrix} = (x,\ y)\mathbf{O}^T(a)\mathbf{O}(a)\begin{bmatrix} \boldsymbol{e}_x \\ \boldsymbol{e}_y \end{bmatrix} = (r,\ 0)\begin{bmatrix} \boldsymbol{e}_r \\ \boldsymbol{e}_a \end{bmatrix} = \boldsymbol{\rho}. \tag{17}$$

And $\|\boldsymbol{\xi}\| = \|\boldsymbol{\rho}\| = r = \|OP\|$, and $\boldsymbol{e}_r = \boldsymbol{e}_x \cos a + \boldsymbol{e}_y \sin a = \boldsymbol{e}_{OP}$.

2.1.1 The Principal Differentials and Jacobian Matrices

Now given $\boldsymbol{\rho} \in \boldsymbol{R}$ and $\boldsymbol{\xi} \in \boldsymbol{X}$ as defined above, when $\boldsymbol{\rho} = \mathbf{h}(\boldsymbol{\xi})$ and $\boldsymbol{\xi} = \mathbf{h}^{-1}(\boldsymbol{\rho})$ their differentials are related as [6]

$$\begin{bmatrix} dr(x,y) \\ da(x,y) \end{bmatrix} = \begin{bmatrix} \frac{\partial}{\partial x} r(x,y) & \frac{\partial}{\partial y} r(x,y) \\ \frac{\partial}{\partial x} a(x,y) & \frac{\partial}{\partial y} a(x,y) \end{bmatrix} \begin{bmatrix} dx \\ dy \end{bmatrix} \tag{18}$$

and

$$\begin{bmatrix} dx(r,a) \\ dy(r,a) \end{bmatrix} = \begin{bmatrix} \frac{\partial}{\partial r} x(r,a) & \frac{\partial}{\partial a} x(r,a) \\ \frac{\partial}{\partial r} y(r,a) & \frac{\partial}{\partial a} y(r,a) \end{bmatrix} \begin{bmatrix} dr \\ da \end{bmatrix}. \tag{19}$$

In the sequel, the above two expressions shall also be written as

$$d\boldsymbol{\rho} = \mathbf{J}(\boldsymbol{\xi})d\boldsymbol{\xi} \text{ and } d\boldsymbol{\xi} = \mathbf{J}^{-1}(\boldsymbol{\rho})d\boldsymbol{\rho}, \tag{20}$$

and the matrices definitized as

$$\mathbf{J}(\boldsymbol{\xi}) = \frac{d\mathbf{h}(\boldsymbol{\xi})}{d\boldsymbol{\xi}^T} \text{ and } \mathbf{J}^{-1}(\boldsymbol{\rho}) = \frac{d\mathbf{h}^{-1}(\boldsymbol{\rho})}{d\boldsymbol{\rho}^T}. \tag{21}$$

The expressions in (20) are linear maps between $\boldsymbol{R}$ and $\boldsymbol{X}$, which have a nonlinear dependency upon a parameter, respectively $\boldsymbol{\xi}$ and $\boldsymbol{\rho}$. The matrices defined by (21) may also be viewed as matrix-valued point functions, written $\mathbf{J}(X)$ and $\mathbf{J}^{-1}(R)$. And when $X = R = P$, $P \neq O$, then $\mathbf{I} = \mathbf{J}(X)\mathbf{J}^{-1}(R) = \mathbf{J}^{-1}(R)\mathbf{J}(X)$.

Of course, $\mathbf{J}$ is the Jacobian matrix of the transformation defined by (12),

$$\mathbf{J}(\boldsymbol{\xi}) \equiv \mathbf{J}(x,y) = \begin{bmatrix} x\big/\sqrt{x^2+y^2} & +y\big/\sqrt{x^2+y^2} \\ -y\big/\left(x^2+y^2\right) & x\big/\left(x^2+y^2\right) \end{bmatrix}, \tag{22}$$

and $\mathbf{J}^{-1}$ is the Jacobian matrix defined by (13),

$$\mathbf{J}^{-1}(\boldsymbol{\rho}) = \mathbf{J}^{-1}(r,a) = \begin{bmatrix} \cos a & -r\sin a \\ +\sin a & r\cos a \end{bmatrix}. \tag{23}$$

And when $\boldsymbol{\rho} = \mathbf{h}(\boldsymbol{\xi})$ and $\boldsymbol{\xi} = \mathbf{h}^{-1}(\boldsymbol{\rho})$, then $\mathbf{J}$ and $\mathbf{J}^{-1}$ may also be written as

$$\mathbf{J}(\boldsymbol{\rho}) = \mathbf{J}(r,a) = \begin{bmatrix} \cos a & +\sin a \\ -(\sin a)/r & (\cos a)/r \end{bmatrix} \tag{24}$$

and

$$\mathbf{J}^{-1}(\boldsymbol{\xi}) \equiv \mathbf{J}^{-1}(x,y) = \begin{bmatrix} x/\sqrt{x^2+y^2} & -y \\ +y/\sqrt{x^2+y^2} & x \end{bmatrix}. \tag{25}$$

Hybrid parameterizations are also defined,

$$\mathbf{J}(x,y;r) = \begin{bmatrix} x/r & +y/r \\ -y/r^2 & x/r^2 \end{bmatrix} \text{ and } \mathbf{J}^{-1}(a;x,y) = \begin{bmatrix} \cos a & -y \\ \sin a & +x \end{bmatrix}. \tag{26}$$

Note that the arguments of $\mathbf{J}$ and $\mathbf{J}^{-1}$ serve two purposes at once: the symbol specifies the parameterization (principal, alternative, or hybrid); and its value specifies the point at which the matrix is being determined. And when $\boldsymbol{\rho}$ and $\boldsymbol{\xi}$ both represent P, then (using the polymorphism of arguments defined above),

$$\mathbf{I} = \mathbf{J}(\boldsymbol{\rho})\mathbf{J}^{-1}(\boldsymbol{\rho}) = \mathbf{J}(\boldsymbol{\xi})\mathbf{J}^{-1}(\boldsymbol{\rho}) = \mathbf{J}(\boldsymbol{\xi})\mathbf{J}^{-1}(\boldsymbol{\xi}) = \mathbf{J}(\boldsymbol{\rho})\mathbf{J}^{-1}(\boldsymbol{\xi}). \tag{27}$$

Finally, $\mathbf{J}$ and $\mathbf{J}^{-1}$ may be factored into products of rotation matrices and metric tensors, written

$$\mathbf{J}(\boldsymbol{\xi}) = \mathbf{D}^{-1}(r)\mathbf{O}(a) \text{ and } \mathbf{J}^{-1}(\mathbf{r}) = \mathbf{O}^T(a)\mathbf{D}(r), \tag{28}$$

with

$$\mathbf{D}(r) = \begin{bmatrix} 1 & 0 \\ 0 & r \end{bmatrix} \text{ and } \mathbf{O}(a) = \begin{bmatrix} \cos a & +\sin a \\ -\sin a & \cos a \end{bmatrix}. \tag{29}$$

Note that $\mathbf{D}^{-1}(r) = \mathbf{D}(1/r)$, $r \neq 0$, and $\mathbf{O}^T(a) = \mathbf{O}^{-1}(a) = \mathbf{O}(-a)$. Also,

$$\|d\mathbf{x}\|^2 = d\mathbf{x}^T d\mathbf{x} = d\mathbf{r}^T\mathbf{D}(r)\mathbf{O}(a)\mathbf{O}^T(a)\mathbf{D}(r)d\mathbf{r} = d\mathbf{r}^T\mathbf{D}^2(r)d\mathbf{r}, \tag{30}$$

which implies

$$\|d\mathbf{x}\|^2 = dx^2 + dy^2 = dr^2 + r^2 da^2 . \tag{31}$$

2.1.2 The Basic Random Vectors

In the sequel, column vectors such as $\boldsymbol{\xi}$ and $\boldsymbol{\rho}$ will be realizations of *random vectors*, denoted $\boldsymbol{X}$ and $\boldsymbol{R}$ [7] – the transposes of $\boldsymbol{X}$ and $\boldsymbol{R}$ shall be written $\boldsymbol{X}^T = (X,\ Y)$ and $\boldsymbol{R}^T = (R,\ A)$ – in this context the symbols X, R, etc. denote scalar random variables, not Euclidean points (the basic notation will be summarized at the end of this Chapter). The expected values of $\boldsymbol{X}$ and $\boldsymbol{R}$ are written

$$\mathcal{E}\boldsymbol{X} = \boldsymbol{\mu}_X = \begin{bmatrix} \mu_X \\ \mu_Y \end{bmatrix} \text{ and } \mathcal{E}\boldsymbol{R} = \boldsymbol{\mu}_R = \begin{bmatrix} \mu_R \\ \mu_A \end{bmatrix},$$

with $\mathcal{E}$ denoting the expectation operator. And the covariance matrix of $\boldsymbol{X}$ is

$$\text{cov}(\boldsymbol{X}) \equiv \mathcal{E}(\boldsymbol{X} - \boldsymbol{\mu}_X)(\boldsymbol{X} - \boldsymbol{\mu}_X)^T = \boldsymbol{\Sigma}_X = \begin{bmatrix} \sigma_X^2 & \sigma_{XY} \\ \sigma_{YX} & \sigma_X^2 \end{bmatrix},$$

with $\sigma_{XY} = \sigma_{YX}$. When Σ_X is positive definite, $\sigma_X^2 > 0$ and $\sigma_X^2 \sigma_Y^2 - \sigma_{XY}^2 > 0$. Similarly,

$$\text{cov}(\boldsymbol{R}) \equiv \mathcal{E}(\boldsymbol{R} - \boldsymbol{\mu}_R)(\boldsymbol{R} - \boldsymbol{\mu}_R)^T = \boldsymbol{\Sigma}_R = \begin{bmatrix} \sigma_R^2 & \sigma_{RA} \\ \sigma_{RA} & \sigma_A^2 \end{bmatrix}.$$

For convenience, a given mean vector and its covariance matrix will be written together, as $(\boldsymbol{\mu}_X;\boldsymbol{\Sigma}_X) \in (\boldsymbol{\mathcal{X}};\boldsymbol{\mathcal{X}}^2)$, where $\boldsymbol{\mu}_X \in \boldsymbol{\mathcal{X}}$ and $\boldsymbol{\Sigma}_X \in \boldsymbol{\mathcal{X}}^2$. And when $\boldsymbol{X}$ and $\boldsymbol{R}$ have gaussian distributions, that shall be indicated by writing $\boldsymbol{X} \sim \mathcal{N}(\boldsymbol{\mu}_X;\boldsymbol{\Sigma}_X)$ and $\boldsymbol{R} \sim \mathcal{N}(\boldsymbol{\mu}_R;\boldsymbol{\Sigma}_R)$. For example, if $\boldsymbol{X} \sim \mathcal{N}(\boldsymbol{\mu}_X;\boldsymbol{\Sigma}_X)$, the density function of $\boldsymbol{X}$ is

$$p_X(\boldsymbol{\xi}) = \frac{1}{2\pi(\det \Sigma_X)^{1/2}} \exp\left[-\frac{1}{2}(\boldsymbol{\xi} - \boldsymbol{\mu}_X)^T \boldsymbol{\Sigma}_X^{-1} (\boldsymbol{\xi} - \boldsymbol{\mu}_X) \right]. \tag{32}$$

Now using the functions $\mathbf{h}$ and $\mathbf{h}^{-1}$ as defined in the previous section, the random vectors $\boldsymbol{X}$ and $\boldsymbol{R}$ may be transformed as $\boldsymbol{R}' \equiv \mathbf{h}(\boldsymbol{X})$ and $\boldsymbol{X}' \equiv \mathbf{h}^{-1}(\boldsymbol{R})$. And given $\boldsymbol{\mu}_X \in \boldsymbol{\mathcal{X}}$ and $\boldsymbol{\mu}_R \in \boldsymbol{\mathcal{R}}$, one can also determine $\mathbf{h}(\boldsymbol{\mu}_X)$ and $\mathbf{h}^{-1}(\boldsymbol{\mu}_R)$. But usually $\boldsymbol{\mu}_{R'} \neq \mathbf{h}(\boldsymbol{\mu}_X)$ and $\boldsymbol{\mu}_{X'} \neq \mathbf{h}^{-1}(\boldsymbol{\mu}_R)$, because $\mathbf{h}$ and $\mathbf{h}^{-1}$ are nonlinear while $\mathcal{E}$ is a linear operator.

Similarly, the covariance matrices of X and R may be transformed as

$$\mathbf{R}'(\xi) = \mathbf{J}(\xi)\Sigma_X\mathbf{J}^T(\xi) \text{ and } \Xi'(\rho) = \mathbf{J}^{-1}(\rho)\Sigma_R\mathbf{J}^{-T}(\rho) \tag{33}$$

($\mathbf{J}^{-T}$ is the transpose of $\mathbf{J}^{-1}$). But usually $\mathbf{R}' \neq \Sigma_{R'}$ and $\Xi' \neq \Sigma_{X'}$ – the transformations in (33) simply represent changes of bases. Of course, under a linear transformation, say $Y = \mathbf{H}X$, where $\mathbf{H}$ is a constant matrix,

$$\mu_Y = \mathcal{E}Y = \mathcal{E}\mathbf{H}X = \mathbf{H}\mathcal{E}X = \mathbf{H}\mu_X \text{ and } \Sigma_Y = \mathcal{E}(Y-\mu_Y)(Y-\mu_Y)^T = \mathbf{H}\Sigma_X\mathbf{H}^T. \tag{34}$$

And, in that case, if $X \sim \mathcal{N}(\mu_X;\Sigma_X)$, then $Y \sim \mathcal{N}(\mathbf{H}\mu_X;\mathbf{H}\Sigma_X\mathbf{H}^T)$ [8].

In the sequel, pairs such as $(\mu_X;\Sigma_X)$ and $(\mu_R;\Sigma_R)$ shall be transformed using $\mathbf{h}$ and $\mathbf{h}^{-1}$ together with the expressions in (33). In particular, two operators are defined, $(\mathbf{h};\mathbf{J})$ and $(\mathbf{h}^{-1};\mathbf{J}^{-1})$, and their respective actions are written succinctly as

$$(\mathbf{h};\mathbf{J}):(\mu_X;\Sigma_X) \mapsto (\rho';\mathbf{R}') \text{ and } (\mathbf{h}^{-1};\mathbf{J}^{-1}):(\mu_R;\Sigma_R) \mapsto (\xi';\Xi'). \tag{35}$$

Note that in (35) the arguments of $\mathbf{h}$ and $\mathbf{h}^{-1}$ are tacitly used as the arguments of $\mathbf{J}$ and $\mathbf{J}^{-1}$. But, under the polymorphism of arguments defined in the previous section for $\mathbf{J}$ and $\mathbf{J}^{-1}$, instead of (33), one may also write

$$\mathbf{R}''(\rho) = \mathbf{J}(\rho)\Sigma_X\mathbf{J}^T(\rho) \text{ and } \Xi''(\xi) = \mathbf{J}^{-1}(\xi)\Sigma_R\mathbf{J}^{-T}(\xi). \tag{36}$$

However, when ξ and ρ represent the same point, say $P \in \mathbb{E}$, $P \neq O$, then $\mathbf{J}(\rho) = \mathbf{J}(\xi)$ and $\mathbf{J}^{-1}(\xi) = \mathbf{J}^{-1}(\rho)$. In which case $\mathbf{R}'(\xi) = \mathbf{R}''(\rho)$ and $\Xi'(\rho) = \Xi''(\xi)$.

Finally, in the sequel, to distinguish realizations of X and R from their deterministic counterparts, measurements of ξ and ρ shall be adorned as $\bar{\xi}$ and $\bar{\rho}$, and estimates adorned as $\hat{\xi}$ and $\hat{\rho}$ (similarly, scalar measurements and estimates will be written $\bar{x}$ and $\hat{x}$, etc.) And measurements and estimates will be written with their covariance matrices as $(\bar{\rho};\Sigma_R)$ and $(\hat{\xi};\Sigma_X)$. However, in radar tracking the underlying random vectors of given measurements and estimates are usually unknown, and surrogates for the true covariance matrices are used instead. Here such surrogates shall be called *associated* covariance matrices, denoted $\mathbf{R}$ and Ξ (and $\mathbf{X}$); and pairs such as $(\bar{\rho};\mathbf{R})$ or $(\hat{\xi};\Xi)$ will be transformed using $(\mathbf{h};\mathbf{J})$ and $(\mathbf{h}^{-1};\mathbf{J}^{-1})$, as in (35).

2.2 Definitions of Detections and Tracks

Here a radar *detection* is assumed to provide an instantaneous measurement on the radar coordinates of some point in $\mathbb{E}$, say $\bar{R}$; and a given detection is assumed to have a certain weight, denoted w, which is based on the signal-to-noise power ratio of the received signal. Also, here a *track* is defined to be a sequence of such detections at distinct times that are believed to be associated with some object. A track that is a single *unassociated* detection, a singlet set, is an *initial-hypothetical* track (if a subsequent detection becomes associated with it, the existence of its object is said to be *verified*). Given its sequence of detections, a track is determined recursively by using a *filter*. A model for the *possible* motion of the object is chosen; and its next detection is *predicted*. And, if a subsequent detection becomes *associated* with that prediction, the track is *updated*. But the association, prediction, and updating operations are determined using either the radar or rectangular coordinate representations (or both). And the more recent detections and the more accurate measurements are given greater weight. Given detections of the same object, different coordinate representations can lead to different tracks. In the next two subsections, the coordinate representations of detections and tracks that shall be used in the sequel are given.

2.2.1 Representations of Detections

Here a detection is defined to be a *weighted point*, denoted $(\bar{R}; w)$, where $\bar{R} \in \mathbb{E}$ is determined by $(\bar{r}, \bar{a})$, and $w > 0$. In $\mathbf{A}$ such is written $(\bar{\rho}; w)$. If the measurements are transformed into rectangular coordinates, using (13), then $\bar{X}$ denotes the point in $\mathbb{E}$, and $\underline{\bar{\xi}}$ is used in $\mathbf{A}$. Of course, $\bar{X} = \bar{R}$ and $\bar{\xi} = \bar{\rho}$, since (13) is exact. And so $(\bar{R}; w) =$ $(\bar{X}; w)$ and $(\bar{\rho}; w) = (\bar{\xi}; w)$ – also, $w\bar{\rho} = w\bar{\xi}$ in $\mathbf{A}$.

Now using the column vector representations of $\bar{R}$ and $\bar{X}$, the basic coordinate forms of a detection are $(\bar{\boldsymbol{\rho}}; \mathbf{I}/w) \in (\boldsymbol{R}; \boldsymbol{R}^2)$ and $(\bar{\boldsymbol{\xi}}; \mathbf{I}/w) \in (\boldsymbol{X}; \boldsymbol{X}^2)$, where $\bar{\boldsymbol{\rho}}^T = (\bar{r}, \bar{a})$ and $\bar{\boldsymbol{\xi}} = \mathbf{h}^{-1}(\bar{\boldsymbol{\rho}})$ – here an inverse weight matrix is being used, $\mathbf{I}/w$. The detection may also be represented by $w\bar{\boldsymbol{\rho}} \in \boldsymbol{R}$ and $w\bar{\boldsymbol{\xi}} \in \boldsymbol{X}$. But, recall, when $\bar{X} = \bar{R}$, then usually $\bar{\boldsymbol{\rho}} \neq \bar{\boldsymbol{\xi}}$ – indeed, when $(\bar{R}; w) = (\bar{X}; w)$, then $w\bar{\boldsymbol{\rho}} \neq w\bar{\boldsymbol{\xi}}$ ($\Pr = 1$).

Range and azimuth measurements, however, usually have disparate accuracies (such are respectively functions of the bandwidth and beamwidth of the radar system [39]). And so, instead of a scalar-weight, a weight-matrix may be better, say

$$\mathbf{W} = \begin{bmatrix} w_r & 0 \\ 0 & w_a \end{bmatrix}, \tag{37}$$

with w_r and w_a both positive. In which case, letting, $\mathbf{R} \equiv \mathbf{W}^{-1}$, the radar detection is written $(\overline{\boldsymbol{\rho}};\mathbf{R})$. In rectangular coordinates such is written $(\overline{\xi};\Xi)$, where $\overline{\xi} = \mathbf{h}^{-1}(\overline{\boldsymbol{\rho}})$ and Ξ = $\mathbf{J}^{-1}(\overline{\boldsymbol{\rho}})\mathbf{R}\mathbf{J}^{-1}(\overline{\boldsymbol{\rho}})$. Note that

$$\mathbf{R}^{-1}\overline{\boldsymbol{\rho}} = \begin{bmatrix} w_r\overline{r} \\ w_a\overline{a} \end{bmatrix} \text{ and } \Xi^{-1}\overline{\xi} = w_r\overline{\xi} = \begin{bmatrix} w_r\overline{x} \\ w_r\overline{y} \end{bmatrix} \tag{38}$$

(see the sequel for details). That is, not only does $\mathbf{R}^{-1}\overline{\boldsymbol{\rho}} \neq \Xi^{-1}\overline{\xi}$ ($\Pr = 1$), but the effective weight of the transformed detection in rectangular coordinates is inconsistent with the given weight of the detection ($\Pr = 1$).

Now if $\overline{\boldsymbol{\rho}}$ is a realization of a known random vector, say $\boldsymbol{R}$, and the covariance matrix of $\boldsymbol{R}$, denoted $\boldsymbol{\Sigma}_{\boldsymbol{R}}$, is also known, then the detection may also be written as $(\overline{\boldsymbol{\rho}};\boldsymbol{\Sigma}_{\boldsymbol{R}})$. Here $\boldsymbol{R} = \boldsymbol{\rho} + \tilde{\boldsymbol{R}}$, where $\boldsymbol{\rho}^T = (r, a)$ is the true position of the object. Usually, $\tilde{\boldsymbol{R}} \sim \mathcal{N}(\mathbf{0};\boldsymbol{\Sigma}_{\boldsymbol{R}})$ is assumed, with

$$\boldsymbol{\Sigma}_{\boldsymbol{R}} = \begin{bmatrix} \sigma_R^2 & 0 \\ 0 & \sigma_A^2 \end{bmatrix}. \tag{39}$$

If $\boldsymbol{\Sigma}_{\boldsymbol{R}}$ is unknown, the detection is written $(\overline{\boldsymbol{\rho}};\mathbf{R})$, with $\mathbf{R}$ some symmetric and positive definite associated covariance matrix. In either case, given $(\overline{\boldsymbol{\rho}};\boldsymbol{\Sigma}_{\boldsymbol{R}})$ or $(\overline{\boldsymbol{\rho}};\mathbf{R})$, the corresponding representation in rectangular coordinates is determined using the second expression in (35), either as

$$(\mathbf{h}^{-1};\mathbf{J}^{-1}):(\overline{\boldsymbol{\rho}};\boldsymbol{\Sigma}_{\boldsymbol{R}}) \mapsto (\overline{\xi};\Xi) \text{ or } (\mathbf{h}^{-1};\mathbf{J}^{-1}):(\overline{\boldsymbol{\rho}};\mathbf{R}) \mapsto (\overline{\xi};\Xi) \tag{40}$$

(the context will specify whether $\boldsymbol{\Sigma}_{\boldsymbol{R}}$ or $\mathbf{R}$ is the pre-image). In the sequel, primes will sometimes be used on these outcomes, as $(\overline{\xi}';\Xi')$, to emphasize that $\overline{\xi}'$ is a *pseudo-measurement*, a realization of some X', and that Ξ' is not the true covariance matrix of X'.

2.2.2 Representations of Tracks

The basic representations of tracks are similar to those defined above for detections: for example, $(\hat{\mathbf{x}};\mathbf{X})$, an estimated vector and its associated covariance (or inverse-weight) matrix. But for a track to be valid, some assumptions on the *possible motion* of the object are also needed. Briefly, the unknown motion of an object is assumed to define a curve in $\mathbb{E}$, a one parameter *continuous* set of points, $\{P(t): t_{min} \le t \le t_{max}\}$, with $t_{min} < t_{max}$ (the independent variable is time). And the curve is assumed to be *sufficiently smooth* (see below). It is said to be degenerate if $P(t) = P(\tau)$, for all τ and t in $[t_{min}, t_{max}]$; and if $P(t) = P(\tau) \Rightarrow t = \tau$, then the curve is said to be *simple* [9].

More formally, an inertial rectangular coordinate frame of reference is invoked [10], namely, $[O;(\boldsymbol{e}_x, \boldsymbol{e}_y)]$. The instantaneous (rectangular) position and velocity vectors with respect to that frame are then ξ and $\dot{\xi}$. And the instantaneous (inertial) acceleration of the object is $\mathbf{a} \equiv \ddot{\xi}$, which is assumed to be piecewise continuous – $\mathbf{a}$ is also assumed to be continuous to the right, and continuous at t_{max}.

Under the assumptions given above for the *possible* motion of the object, when $\mathbf{a}$ is independent of ξ and $\dot{\xi}$, the equation of the *actual* motion is [50]

$$\frac{d}{dt}\begin{bmatrix}\xi \\ \dot{\xi}\end{bmatrix} = \begin{bmatrix}0 & 1 \\ 0 & 0\end{bmatrix}\begin{bmatrix}\xi \\ \dot{\xi}\end{bmatrix} + \begin{bmatrix}0 \\ \mathbf{a}\end{bmatrix}. \tag{41}$$

And so, letting $\xi_0 = \xi(t_0)$ and $\dot{\xi}_0 = \dot{\xi}(t_0)$ at t_0, the position and motion of the object at t are [51]

$$\begin{bmatrix}\xi(t) \\ \dot{\xi}(t)\end{bmatrix} = \mathbf{\Phi}(t_0, t)\begin{bmatrix}\xi_0 \\ \dot{\xi}_0\end{bmatrix} + \int_{t_0}^{t} \mathbf{\Phi}(\tau, t)\begin{bmatrix}0 \\ \mathbf{a}(\tau)\end{bmatrix} d\tau, \tag{42}$$

where $\mathbf{\Phi}$ is the one-sided Green's function matrix associated with the matrix in (41). Recall that the fundamental property of $\mathbf{\Phi}$ is $\mathbf{\Phi}(a,c) = \mathbf{\Phi}(a,b)\mathbf{\Phi}(b,c)$, and that $\mathbf{\Phi}(t,t) =$ I and $\mathbf{\Phi}^{-1}(\tau,t) = \mathbf{\Phi}(t,\tau)$. In particular, letting

$$\mathbf{F} = \begin{bmatrix}0 & 1 \\ 0 & 0\end{bmatrix}, \tag{43}$$

$\mathbf{\Phi}(\tau,t) = e^{(t-\tau)\mathbf{F}}$. In which case,

$$\mathbf{\Phi}(t-\tau) \equiv \mathbf{\Phi}(\tau,t) = \begin{bmatrix}1 & t-\tau \\ 0 & 1\end{bmatrix}. \tag{44}$$

Of course, in radar tracking ξ and $\dot{\xi}$ are generally unknown, and $\mathbf{a}$ is also unknown (and the times at which $\mathbf{a}$ is discontinuous are also unknown).

Now the basic filtering problem of radar tracking is: given a sequence of radar detections on the object, at times t_k, $k = 0,1,\cdots,K$ (henceforth, K shall be used exclusively to denote certain integers), ordered as $t_m < t_n$ when $m<n$, recursively determine estimates of $\xi_k \equiv \xi(t_k)$ and $\dot{\xi}_k \equiv \dot{\xi}(t_k)$. If the object is believed to be maneuvering, $\ddot{\xi}_k = \ddot{\xi}(t_k)$ may also need to be estimated; and if the object is believed to be motionless (the degenerate case), only ξ_k needs to be estimated.

Now let $\mathbf{x}^T = (\xi^T, \dot{\xi}^T)$, and let $\mathbf{f}^T = (\mathbf{0}^T, \mathbf{a}^T)$. In which case, (41) becomes

$$\dot{\mathbf{x}} = \mathbf{F}\mathbf{x} + \mathbf{f}, \ \mathbf{x}_0 = \mathbf{x}(t_0), \tag{45}$$

and the solution is

$$\mathbf{x}(t) = \mathbf{\Phi}(t-t_0)\mathbf{x}_0 + \int_{t_0}^{t} \mathbf{\Phi}(t-\tau)\mathbf{f}(\tau)d\tau \,. \tag{46}$$

Accordingly, the track is denoted by $(\hat{\mathbf{x}};\mathbf{X})$, with $\mathbf{X}$ some associated covariance matrix of $\hat{\mathbf{x}}$ (the determination of $\mathbf{X}$ is discussed in the next section). In this representation, the elements of the track are respectively

$$\hat{\mathbf{x}} = \begin{bmatrix} \hat{\xi} \\ \hat{\dot{\xi}} \end{bmatrix} \text{ and } \mathbf{X} = \begin{bmatrix} \Xi_{\xi} & \Xi_{\xi\dot{\xi}} \\ \Xi_{\dot{\xi}\xi} & \Xi_{\dot{\xi}} \end{bmatrix}, \tag{47}$$

with $\mathbf{X}$ tacitly assumed to be symmetric and positive definite (Ξ_{ξ} and $\Xi_{\dot{\xi}}$ symmetric and positive definite, and $\Xi_{\dot{\xi}\xi} = \Xi_{\xi\dot{\xi}}^T$). When the object is assumed to be motionless or maneuvering, the track will still be denoted by $(\hat{\mathbf{x}};\mathbf{X})$, but in those cases the elements are

$$\hat{\mathbf{x}} = \hat{\xi} \text{ and } \mathbf{X} = \Xi_{\xi}, \tag{48}$$

and

$$\hat{\mathbf{x}} = \begin{bmatrix} \hat{\xi} \\ \hat{\dot{\xi}} \\ \hat{\ddot{\xi}} \end{bmatrix} \text{ and } \mathbf{X} = \begin{bmatrix} \Xi_{\xi} & \Xi_{\xi\dot{\xi}} & \Xi_{\xi\ddot{\xi}} \\ \Xi_{\dot{\xi}\xi} & \Xi_{\dot{\xi}} & \Xi_{\dot{\xi}\ddot{\xi}} \\ \Xi_{\ddot{\xi}\xi} & \Xi_{\ddot{\xi}\dot{\xi}} & \Xi_{\ddot{\xi}} \end{bmatrix}. \tag{49}$$

These three cases are usually called *constant position* (CP), (48), *constant velocity* (CV), (47), and *constant acceleration* (CA), (49). Note that, in the CP case,

$$\mathbf{F} = 1 \quad \text{and} \quad \mathbf{\Phi}(t-\tau) = 1, \tag{50}$$

the CV case has (43) and (44), and in the CA case

$$\mathbf{F} = \begin{bmatrix} 0 & 1 & 0 \\ 0 & 0 & 1 \\ 0 & 0 & 0 \end{bmatrix} \quad \text{and} \quad \mathbf{\Phi}(t-\tau) = \begin{bmatrix} 1 & t-\tau & (t-\tau)^2/2 \\ 0 & 1 & t-\tau \\ 0 & 0 & 1 \end{bmatrix}. \tag{51}$$

Radar coordinates may also be used to denote a radar track, $(\hat{\mathbf{r}};\mathbf{R})$, with $\hat{\mathbf{r}}$ an estimate of $\mathbf{r}$, and $\mathbf{R}$ the associated covariance matrix of $\hat{\mathbf{r}}$. And the CV model may be used in radar coordinates, $\dot{\mathbf{r}} = \mathbf{F}\mathbf{r}$ with $\mathbf{r}^T = (\boldsymbol{\rho}^T, \dot{\boldsymbol{\rho}}^T)$. But such defines a spiral in $\mathbb{E}$; and, in that case, CP, CV, CA are misnomers – more formally, they denote the (kinematic) order of the model, $CP \equiv 0$, $CV \equiv 1$, $CA \equiv 2$. Note that the same symbol may be used for the associated covariances matrices of detections and tracks. In the sequel, such will be distinguished by writing $(\bar{\boldsymbol{\rho}};\bar{\mathbf{R}})$ and $(\hat{\mathbf{r}};\hat{\mathbf{R}})$ when the context requires. And $(\bar{\boldsymbol{\xi}};\bar{\mathbf{X}})$ and $(\hat{\mathbf{x}};\hat{\mathbf{X}})$ will also be used when needed.

Finally, tracks in rectangular coordinates are transformed into radar coordinates, and vice versa. The position sub-vectors are transformed using (12) and (14), and the velocity and acceleration sub-vectors transformed using the first and second derivatives of those functions. For example, a rectangular CV track, $(\hat{\mathbf{x}};\hat{\mathbf{X}})$, where $\hat{\mathbf{x}}^T = (\hat{\boldsymbol{\xi}}^T, \hat{\dot{\boldsymbol{\xi}}}^T)$, is transformed into a radar CP track, $(\hat{\boldsymbol{\rho}};\hat{\mathbf{R}})$, by using $(\mathbf{h};\mathbf{H})$, with $\boldsymbol{\rho} = \mathbf{h}(\boldsymbol{\xi})$ and $\mathbf{H} = [\mathbf{J} \quad \mathbf{0}]$ – in the CP and CA cases, respectively, the corresponding covariance matrix transformation uses $\mathbf{H} = \mathbf{J}$ and $\mathbf{H} = [\mathbf{J} \quad \mathbf{0} \quad \mathbf{0}]$. (The context shall specify the details.)

2.3 The Basic Tracking Equations

In this section the *linear Kalman filter* (LKF) equations that are commonly used in radar tracking are provided – such are taken from [11]. To expedite the presentation, the form of the detection shall be $(\bar{\mathbf{y}};\mathbf{Y})$, where $\bar{\mathbf{y}}$ is a measurement of $\mathbf{y}$, with $\mathbf{y} = \mathbf{H}\mathbf{x}$ (a *linear measurement model* is being used for now). And so here $(\mathbf{H};\mathbf{H}):(\mathbf{x};\mathbf{X}) \mapsto (\mathbf{y};\mathbf{Y})$. First, the basic tracking algorithm is outlined, and then the details are provided.

Given the first few detections that are *believed* to be associated with a new object, a "batch" estimator (weighted least-squares, or the BLUE) is used to determine its *initial* track, denoted $(\hat{\mathbf{x}}_0;\hat{\mathbf{X}}_0)$ at t_0. The subsequent detections are then reindexed as t_n, $n = 1, 2, \cdots, N$. Next, given $(\hat{\mathbf{x}}_{n-1};\hat{\mathbf{X}}_{n-1})$ at t_{n-1}, a model of motion (an assumption) is used to predict the track at t_n, denoted $(\hat{\mathbf{x}}_n^-;\hat{\mathbf{X}}_n^-)$. And a predicted detection at t_n is determined,

$$(\mathbf{H};\mathbf{H}):(\hat{\mathbf{x}}_n^-;\hat{\mathbf{X}}_n^-)\mapsto(\hat{\mathbf{y}}_n^-;\hat{\mathbf{Y}}_n^-). \tag{52}$$

In practice, a hypothesis test [12] may also be needed to choose the next detection – something like the Mahalanobis distance [13], with some decision threshold γ,

$$(\hat{\mathbf{y}}_n^- - \bar{\mathbf{y}}_n)^T\left(\hat{\mathbf{Y}}_n^- + \bar{\mathbf{Y}}_n\right)^{-1}(\hat{\mathbf{y}}_n^- - \bar{\mathbf{y}}_n)^T \underset{\mathcal{H}_A}{\overset{\mathcal{H}_U}{\gtreqless}} \gamma^2, \tag{53}$$

where $\mathcal{H}_A$ is the null-hypothesis (*associated*) and $\mathcal{H}_U$ is the alternative one (*unassociated*). Under $\mathcal{H}_A$, $(\bar{\mathbf{y}}_n;\bar{\mathbf{Y}}_n)$ is used to *update* $(\hat{\mathbf{x}}_n^-;\hat{\mathbf{X}}_n^-)$, which yields $(\hat{\mathbf{x}}_n;\hat{\mathbf{X}}_n)$.

2.3.1 The Prediction Step

In the Kalman approach to tracking the *possible* positions and motions of the object are modeled as a mean-squared continuous random process, having the form [14]

$$\frac{d}{dt}\boldsymbol{X}_t = \mathbf{F}\boldsymbol{X}_t + \boldsymbol{V}_t,\ \boldsymbol{X}_0 = \boldsymbol{X}(0) \tag{54}$$

(for convenience, $0 \le t \le T$, and the initial conditions are given at $t_0 = 0$). The above equation is called the *system model*, and the random process $\boldsymbol{V}_t$ is called the *system noise*, assumed to satisfy $\mathcal{E}\boldsymbol{V}_t = \mathbf{0}$, with $\mathcal{E}\boldsymbol{V}_t\boldsymbol{V}_\tau = \mathbf{S}_V(t,\tau)\delta(t-\tau)$, where $\delta(t-\tau)$ is the Dirac delta function. Also, $\boldsymbol{V}_t$ is assumed to be independent of $\boldsymbol{X}_0$ for all $t \in (0,1]$, with $\mathbf{S}_V(t)=$ $\mathbf{S}_V(t,t)$ positive semi-definite (in the literature $\mathbf{S}(t)$ is called the spectral density matrix of the system noise). Note that (54) is analogous to the deterministic case,

$$\dot{\mathbf{x}}(t) = \mathbf{F}\mathbf{x}(t) + \mathbf{f}(t),\ \mathbf{x}(0) = \mathbf{x}_0. \tag{55}$$

Now under the above assumptions,

$$\dot{\boldsymbol{\mu}}_X(t) = \mathbf{F}\boldsymbol{\mu}_X(t) \text{ and } \dot{\boldsymbol{\Sigma}}_X(t) = \mathbf{F}\boldsymbol{\Sigma}_X(t) + \boldsymbol{\Sigma}_X(t)\mathbf{F}^T + \mathbf{S}(t) \tag{56}$$

And so,

$$\boldsymbol{\mu}_X(t) = \Phi(t)\boldsymbol{\mu}_X(0) \text{ and } \boldsymbol{\Sigma}_X(t) = \Phi(t)\boldsymbol{\Sigma}_X(0)\Phi^T(t) + \boldsymbol{\Sigma}_V(t) \tag{57}$$

where

$$\mathbf{\Sigma}_V(t) \equiv \int_0^t \Phi(t-\tau)\mathbf{S}(\tau)\Phi^T(t-\tau)d\tau . \tag{58}$$

Such defines the prediction step. For convenience, it shall be written symbolically as

$$(\boldsymbol{\mu}_X;\mathbf{\Sigma}_X) \xrightarrow{(\mathbf{F};\mathbf{S})} (\hat{\mathbf{x}}^-;\hat{\mathbf{X}}^-), \tag{59}$$

with the context providing the necessary details (i.e., the kinematic order, number of geometric degrees-of-freedom, and the coordinate representation).

As an example, consider the one degree-of-freedom rectangular *CV* case. Let $\mathbf{x}^T = (x,\ \dot{x})$; and let the initial conditions to (56) be

$$\hat{\mathbf{x}}_0 = \begin{bmatrix} \mu_X \\ \mu_{\dot{X}} \end{bmatrix} \text{ and } \hat{\mathbf{X}}_0 = \begin{bmatrix} \sigma_X^2 & \sigma_{X\dot{X}} \\ \sigma_{X\dot{X}} & \sigma_{\dot{X}}^2 \end{bmatrix}. \tag{60}$$

And let

$$\mathbf{S}(t) = \begin{bmatrix} 0 & 0 \\ 0 & s \end{bmatrix}, \tag{61}$$

with s a non-negative constant. The prediction at time t is then

$$\hat{\mathbf{x}}(t^-) = \Phi(t)\hat{\mathbf{x}}_0 = \begin{bmatrix} \mu_X + t\mu_{\dot{X}} \\ \mu_{\dot{X}} \end{bmatrix} \tag{62}$$

and

$$\hat{\mathbf{X}}(t^-) = \Phi(t)\hat{\mathbf{X}}_0\Phi^T(t) + \mathbf{\Sigma}_V(t), \tag{63}$$

with

$$\Phi(t)\hat{\mathbf{X}}_0\Phi^T(t) = \begin{bmatrix} \sigma_X^2 + 2t\sigma_{X\dot{X}} + t^2\sigma_{\dot{X}}^2 & \sigma_{X\dot{X}} + t\sigma_{\dot{X}}^2 \\ \sigma_{X\dot{X}} + t\sigma_{\dot{X}}^2 & \sigma_{\dot{X}}^2 \end{bmatrix} \tag{64}$$

and

$$\boldsymbol{\Sigma}_V(t) = s\begin{bmatrix} t^3/3 & t^2/2 \\ t^2/2 & t \end{bmatrix}. \tag{65}$$

2.3.2 The Update Step

The estimation equations of the Kalman filter are basically those of the Best Linear Unbiased Estimator (BLUE) – a derivation of such, also taken from [11], is given in the Appendix. The basic assumptions are that the underlying random vectors of the detection's measurement and the predicted track are mutually independent.

Given a predicted track and measurement at time t, respectively $(\hat{\mathbf{x}}^-;\hat{\mathbf{X}}^-)$ and $(\bar{\mathbf{y}};\bar{\mathbf{Y}})$, with $\bar{\mathbf{Y}}$ and $\hat{\mathbf{X}}^-$ positive definite, the standard Kalman update equations are

$$\hat{\mathbf{x}} = \hat{\mathbf{x}}^- + \mathbf{K}\left(\bar{\mathbf{y}} - \mathbf{H}\hat{\mathbf{x}}^-\right) \text{ and } \hat{\mathbf{X}} = \left(\mathbf{I} - \mathbf{K}\mathbf{H}\right)\hat{\mathbf{X}}^-, \tag{66}$$

where

$$\mathbf{K} = \hat{\mathbf{X}}^-\mathbf{H}^T\left(\mathbf{H}\hat{\mathbf{X}}^-\mathbf{H}^T + \bar{\mathbf{Y}}\right)^{-1}. \tag{67}$$

An alternate form is [15]

$$\hat{\mathbf{x}} = \hat{\mathbf{X}}\left[(\hat{\mathbf{X}}^-)^{-1}\hat{\mathbf{x}}^- + \mathbf{H}^T\bar{\mathbf{Y}}^{-1}\bar{\mathbf{y}}\right] \text{ and } \hat{\mathbf{X}}^{-1} = (\hat{\mathbf{X}}^-)^{-1} + \mathbf{H}^T\bar{\mathbf{Y}}^{-1}\mathbf{H}. \tag{68}$$

with

$$\mathbf{K} = \hat{\mathbf{X}}\mathbf{H}^T\bar{\mathbf{Y}}^{-1}. \tag{69}$$

Note that, using (69) in the first expression of (66), and letting $\hat{\mathbf{y}}^- = \mathbf{H}\hat{\mathbf{x}}^-$, leads to

$$\hat{\mathbf{x}} = \hat{\mathbf{x}}^- + \hat{\mathbf{X}}\mathbf{H}^T\bar{\mathbf{Y}}^{-1}\left(\bar{\mathbf{y}} - \hat{\mathbf{y}}^-\right). \tag{70}$$

This alternate form is more amenable to analysis. For example, the difference $\bar{\mathbf{y}} - \hat{\mathbf{y}}^-$ is called the (prediction) *residual*; and $\bar{\mathbf{Y}}^{-1}(\bar{\mathbf{y}} - \hat{\mathbf{y}}^-)$ and $\hat{\mathbf{X}}^{-1}(\hat{\mathbf{x}} - \hat{\mathbf{x}}^-)$ are called (normalized) *innovations*. Using (70), the innovations are readily seen to be related as

$$\hat{\mathbf{X}}^{-1}\left(\hat{\mathbf{x}} - \hat{\mathbf{x}}^-\right) = \mathbf{H}^T\bar{\mathbf{Y}}^{-1}\left(\bar{\mathbf{y}} - \hat{\mathbf{y}}^-\right), \tag{71}$$

which has the form of a differential. For convenience, the update shall be written symbolically as

$$(\hat{\mathbf{x}}^-;\hat{\mathbf{X}}^-)\xrightarrow{(\bar{\mathbf{y}};\bar{\mathbf{Y}})}(\hat{\mathbf{x}};\hat{\mathbf{X}}), \tag{72}$$

with the details provided by the context.

As an example, consider the one degree-of-freedom rectangular *CV* case, where $\mathbf{x}^T=(x,\ \dot{x})$. Here the predicted track is $(\hat{\mathbf{x}}^-;\hat{\mathbf{X}}^-)$. In component form,

$$\hat{\mathbf{x}}^-=\begin{bmatrix}\hat{x}^-\\ \hat{\dot{x}}^-\end{bmatrix} \text{ and } \hat{\mathbf{X}}^-=\begin{bmatrix}(\sigma_{XX}^-)^2 & (\sigma_{X\dot{X}}^-)^2\\ (\sigma_{\dot{X}X}^-)^2 & (\sigma_{\dot{X}\dot{X}}^-)^2\end{bmatrix}. \tag{73}$$

And the "detection" is simply $(\bar{y};\sigma_Y^2)$, where $\bar{y}=\mathbf{H}\mathbf{x}+\tilde{y}$ with $\mathbf{H}=(1,\ 0)$. In which case, using (70), the updated estimate is $(\hat{\mathbf{x}};\hat{\mathbf{X}})$, where

$$\hat{\mathbf{x}}=\begin{bmatrix}\hat{x}\\ \hat{\dot{x}}\end{bmatrix}=\begin{bmatrix}\hat{x}^-\\ \hat{\dot{x}}^-\end{bmatrix}+\begin{bmatrix}\sigma_{XX}^2/\sigma_Y^2\\ \sigma_{\dot{X}X}^2/\sigma_Y^2\end{bmatrix}(\bar{y}-\hat{y}^-) \tag{74}$$

and

$$\hat{\mathbf{X}}^{-1}=\begin{bmatrix}\sigma_{XX}^2 & \sigma_{X\dot{X}}^2\\ \sigma_{\dot{X}X}^2 & \sigma_{\dot{X}\dot{X}}^2\end{bmatrix}^{-1}=\begin{bmatrix}(\sigma_{XX}^-)^2 & (\sigma_{X\dot{X}}^-)^2\\ (\sigma_{\dot{X}X}^-)^2 & (\sigma_{\dot{X}\dot{X}}^-)^2\end{bmatrix}^{-1}+\begin{bmatrix}1/\sigma_Y^2 & 0\\ 0 & 0\end{bmatrix}. \tag{75}$$

Here the gain matrix, (69), is

$$\mathbf{K}=\frac{1}{(\sigma_{XX}^-)^2+\sigma_Y^2}\begin{bmatrix}(\sigma_{XX}^-)^2\\ (\sigma_{\dot{X}X}^-)^2\end{bmatrix}=\begin{bmatrix}\sigma_{XX}^2/\sigma_Y^2\\ \sigma_{\dot{X}X}^2/\sigma_Y^2\end{bmatrix}. \tag{76}$$

2.4 A Special Case

In the sequel, when evaluating the results of the analyses, certain closed form solutions to the tracking problem will be used as accuracy references. In this section those references are derived (such are the Cramer-Rao lower bounds of the linear-gaussian case having zero system noise). In particular, let $\mathbf{x}^T=(x,\ \dot{x})$, with $\ddot{x}=0$ and $s=0$. And let the detections provide measurements directly on x, written $(\bar{x}_n;\sigma_X^2)$, and given at equispaced intervals of time, $\tau \equiv t_n - t_{n-1}$, $n = 2,3,\cdots,N$. In which case the propagation and update equations can be combined into a single "batch" form as follows.

First, define $\bar{\mathbf{y}}^T \equiv (\bar{x}_1, \bar{x}_2, \cdots, \bar{x}_N)$ and $\mathbf{\Sigma}_Y \equiv \sigma_{\tilde{X}}^2 \mathbf{I}$ (the random vectors of the measurement errors are orthogonal). And define the "batch detection" to be $(\bar{\mathbf{y}};\mathbf{\Sigma}_Y)$, along with the batch measurement model, $\bar{\mathbf{y}} = \mathbf{H}\mathbf{x}(t_N) + \tilde{\mathbf{y}}$, where

$$\mathbf{H}^T \equiv \begin{bmatrix} 1 & \cdots & 1 & 1 \\ -(N-1)\tau & \cdots & -\tau & 0 \end{bmatrix}. \tag{77}$$

Then, given $(\bar{\mathbf{y}};\mathbf{\Sigma}_Y)$ with $N \geq 2$, the (batch) BLUE equations are

$$\hat{\mathbf{x}}(t_N) = \hat{\mathbf{X}}(t_N)\mathbf{H}^T\mathbf{\Sigma}_Y^{-1}\bar{\mathbf{y}} \quad \text{and} \quad \hat{\mathbf{X}}(t_N) = (\mathbf{H}^T\mathbf{\Sigma}_Y^{-1}\mathbf{H})^{-1}. \tag{78}$$

But, since $\mathbf{\Sigma}_Y = \sigma_{\tilde{X}}^2 \mathbf{I}$, this simplifies to

$$\hat{\mathbf{x}}(t_N) = (\mathbf{H}^T\mathbf{H})^{-1}\mathbf{H}^T\bar{\mathbf{y}} \quad \text{and} \quad \hat{\mathbf{X}}(t_N) = \sigma_{\tilde{X}}^2(\mathbf{H}^T\mathbf{H})^{-1}, \tag{79}$$

with

$$\mathbf{H}^T\mathbf{H} = \sum_{n=1}^{N} \begin{bmatrix} 1 & -(n-1)\tau \\ -(n-1)\tau & (n-1)^2\tau^2 \end{bmatrix}. \tag{80}$$

Of course,

$$\sum_{n=0}^{N-1} 1 = N, \quad \sum_{n=1}^{N-1} n = \frac{(N-1)N}{2}, \quad \sum_{n=0}^{N-1} n^2 = \frac{(N-1)N(2N-1)}{6}.$$

And so

$$\mathbf{H}^T\mathbf{H} = \begin{bmatrix} N & -(N-1)N\tau/2 \\ -(N-1)N\tau/2 & (N-1)N(2N-1)\tau^2/6 \end{bmatrix} \tag{81}$$

and

$$\det \mathbf{H}^T\mathbf{H} = (N-1)N^2(N+1)\tau^2/12. \tag{82}$$

Thus, for $N \geq 2$,

$$\hat{\mathbf{X}}(t_N) = \sigma_X^2 \begin{bmatrix} 2(2N-1)/N(N+1) & 6/N(N+1)\tau \\ 6/N(N+1)\tau & 12/(N^2-1)N\tau^2 \end{bmatrix}. \tag{83}$$

Now, if a subsequent detection is given at t_{N+1}, the prediction $(\hat{\mathbf{x}}_{N+1}^-;\hat{\mathbf{X}}_{N+1}^-)$ may be updated using $(\overline{x}_{N+1};\sigma_X^2)$, to determine $(\hat{\mathbf{x}}_{N+1};\hat{\mathbf{X}}_{N+1})$. In particular,

$$\hat{\mathbf{x}}_{N+1} = \hat{\mathbf{x}}_{N+1}^- + \mathbf{K}_{N+1}(\overline{x}_{N+1} - \hat{x}_{N+1}^-) \text{ and } \hat{\mathbf{X}}_{N+1} = (\mathbf{I} - \mathbf{K}_{N+1}\mathbf{G})\hat{\mathbf{X}}_{N+1}^-, \tag{84}$$

with $\mathbf{G} = [1 \quad 0]$ and

$$\mathbf{K}_{N+1} = \hat{\mathbf{X}}_{N+1}\mathbf{G}^T\sigma_X^{-2} = \begin{bmatrix} \alpha_{N+1} \\ \beta_{N+1}/\tau \end{bmatrix}, \tag{85}$$

where, from (83) and the definition of $\mathbf{G}$,

$$\alpha_N = 2(2m-1)/(m^2+m) \text{ and } \beta_N = 6/(m^2+m), \tag{86}$$

with $m = N+2$, $N > 0$. (This special case is the theoretical basis for the so-called *alpha-beta* filter, which is perhaps the most commonly used radar tracking method [16].)

Note that the variance of the position estimate in (83) is asymptotically $4\sigma_X^2/N$. In the *CP* case it is exactly σ_X^2/N. And in the CA case it is asymptotically $9\sigma_X^2/N$. Indeed, if a p -th kinematic order model were used, satisfying the above assumptions, the variance of the position estimate would be asymptotically $\sigma_X^2(p+1)^2/N$ [17, 18].

2.5 The Basic Estimation Cases and Notation

In the sequel several different estimators shall be used. For convenience, the basic ones are summarized here (with time not explicitly denoted) – the next Chapter shall provide the details. All are predictor-corrector loops as defined above: given a prior track, the propagation equations are used to determine a prediction; and, given a predicted track and a detection, the update equations are used to determine a correction.

- The Linear Kalman Filter (LKF). Here the radar detections are used directly to determine the track in radar coordinates. The prediction and update steps are simply

$$(\hat{\mathbf{r}};\hat{\mathbf{R}}) \xrightarrow{(\mathbf{F};\mathbf{S})} (\hat{\mathbf{r}}^-;\hat{\mathbf{R}}^-) \text{ and } (\hat{\mathbf{r}}^-;\hat{\mathbf{R}}^-) \xrightarrow{(\mathbf{r};\overline{\mathbf{R}})} (\hat{\mathbf{r}};\hat{\mathbf{R}}). \tag{87}$$

- The Pseudo-LKF (PLKF). This case uses the LKF, but with the detections first converted into rectangular coordinates. That is, given $(\bar{\mathbf{r}};\bar{\mathbf{R}})$, first determine

$$(\mathbf{h}^{-1};\mathbf{J}^{-1}):(\bar{\mathbf{r}};\bar{\mathbf{R}})\mapsto(\bar{\mathbf{x}}';\bar{\mathbf{X}}'), \tag{88}$$

called a pseudo-detection. And then use

$$(\hat{\mathbf{x}};\hat{\mathbf{X}})\xrightarrow{(\mathbf{F};\mathbf{S})}(\hat{\mathbf{x}}^{-};\hat{\mathbf{X}}^{-}) \text{ and } (\hat{\mathbf{x}}^{-};\hat{\mathbf{X}}^{-})\xrightarrow{(\bar{\mathbf{x}}';\bar{\mathbf{X}}')}(\hat{\mathbf{x}};\hat{\mathbf{X}}). \tag{89}$$

Alternatively, the update step may be written

$$(\hat{\mathbf{x}}^{-};\hat{\mathbf{X}}^{-})\xrightarrow{(\mathbf{h}^{-1};\mathbf{J}^{-1}):(\bar{\mathbf{r}};\bar{\mathbf{R}})\mapsto(\bar{\mathbf{x}}';\bar{\mathbf{X}}')}(\hat{\mathbf{x}};\hat{\mathbf{X}}). \tag{90}$$

- The Converted-LKF (CLKF). Here the track is predicted in rectangular coordinates and updated in radar coordinates. In particular, before and after each update the track is respectively transformed as

$$(\mathbf{h};\mathbf{H}):(\hat{\mathbf{x}}^{-};\hat{\mathbf{X}}^{-})\mapsto(\hat{\mathbf{r}}^{-};\hat{\mathbf{R}}^{-}) \text{ and } (\mathbf{h}^{-1};\mathbf{H}^{-1}):(\hat{\mathbf{r}};\hat{\mathbf{R}})\mapsto(\hat{\mathbf{x}};\hat{\mathbf{X}}). \tag{91}$$

And the prediction and update steps are

$$(\hat{\mathbf{x}};\hat{\mathbf{X}})\xrightarrow{(\mathbf{F};\mathbf{S})}(\hat{\mathbf{x}}^{-};\hat{\mathbf{X}}^{-}) \text{ and } (\hat{\mathbf{r}}^{-};\hat{\mathbf{R}}^{-})\xrightarrow{(\mathbf{r};\bar{\mathbf{R}})}(\hat{\mathbf{r}};\hat{\mathbf{R}}). \tag{92}$$

A variation of this case is the Radar Principal Cartesian Coordinates (RPCC) method [58]. There the prediction update steps are

$$(\hat{\mathbf{x}};\hat{\mathbf{R}})\xrightarrow{(\mathbf{F};\mathbf{S})}(\hat{\mathbf{x}}^{-};\hat{\mathbf{R}}^{-}) \text{ and } (\hat{\mathbf{r}}^{-};\hat{\mathbf{R}}^{-})\xrightarrow{(\mathbf{r};\bar{\mathbf{R}})}(\hat{\mathbf{r}};\hat{\mathbf{R}}), \tag{93}$$

with $(\mathbf{h};\mathbf{I}):(\hat{\mathbf{x}}^{-};\hat{\mathbf{R}}^{-})\mapsto(\hat{\mathbf{r}}^{-};\hat{\mathbf{R}}^{-})$ and $(\mathbf{h}^{-1};\mathbf{I}):(\hat{\mathbf{r}};\hat{\mathbf{R}})\mapsto(\hat{\mathbf{x}};\hat{\mathbf{R}})$.

- The Extended Kalman Filter (EKF). This case uses the radar detections directly to update the track in rectangular coordinates. Its prediction and upstate steps are

$$(\hat{\mathbf{x}};\hat{\mathbf{X}})\xrightarrow{(\mathbf{F};\mathbf{S})}(\hat{\mathbf{x}}^{-};\hat{\mathbf{X}}^{-}) \text{ and } (\hat{\mathbf{x}}^{-};\hat{\mathbf{X}}^{-})\xrightarrow{(\mathbf{r};\bar{\mathbf{R}})}(\hat{\mathbf{x}};\hat{\mathbf{X}}). \tag{94}$$

Finally, the basic notation is summarized as follows. Non-bold italic lowercase symbols shall denote real scalars; and non-bold italic uppercase symbols shall denote either Euclidean points or random variables (the context will specify). Bold symbols such as ξ, $\mathbf{x}$, X, $\mathbf{X}$ shall denote objects that are referenced to rectangular coordinates; while ρ, $\mathbf{r}$, R, $\mathbf{R}$ shall denote the corresponding ones that are referenced to radar coordinates. In particular, column vectors of scalars will be denoted by symbols such as $\mathbf{x}$ and $\mathbf{r}$; when they are realizations of random vectors, those functions will be denoted as X and R; and symbols such as $\mathbf{X}$ and $\mathbf{R}$ will denote the corresponding covariance matrices. There will be exceptions to these rules: for example, the domain of the independent variable, time, a closed real interval, written $[0,T]$, with "T" a real number; X,Y,R,A may also denote random variables (the context shall specify); and K,M,N,L shall always denote integers.

2.6 Chapter 2 References

[1] D. Hilbert, Foundations of Geometry, translated D. Bernays 1971, Open Court (1996).

[2] G. A. Jennings, Modern Geometry with Applications, Springer (1994).

[3] J. W. Young and W. W. Denton, Fundamental Concepts of Algebra and Geometry, MacMillan (1930).

[4] J. Stillwell, The Four Pillars of Geometry, Springer Science, (2005).

[5] J. Gallier, Geometric Methods and Applications for Computer Science and Engineering, Springer-Verlag (2001).

[6] T. M. Apostol, Mathematical Analysis, Addison-Wesley (1974).

[7] K. S. Miller, Engineering Mathematics, Dover (1963).

[8] T. W. Anderson, An Introduction to Multivariate Statistical Analysis, Wiley (2003).

[9] E. Kreyszig, Differential Geometry, Dover (1991).

[10] G. Blass, Theoretical Physics, Appleton-Century-Crofts, (1962).

[11] K. S. Miller and D. M. Leskiw, An Introduction to Kalman Filtering With Applications, Krieger (1987).

[12] S. M. Kay, Fundamentals of Statistical Signal Processing, Volume 2: Detection Theory, Prentice Hall (1998)

[13] P. C. Mahalanobis, "On the Generalized Distance in Statistics," in Proceedings of the National Institute of Sciences of India 2 (1), pp. 49-55 (1938).

[14] K. S. Miller, An Introduction to Vector Stochastic Processes, Krieger (1980).

[15] A. Gelb (Ed), Applied Optimal Estimation, MIT Press (1974).

[16] E. Brookner, Tracking and Kalman Filtering Made Easy, Wiley (1998).

[17] D. M. Leskiw and K. S. Miller, "Convergence Properties in Kalman Filtering,"

Journal of Information and Optimization Sciences, Vol. 1, No. 3, pp. 197-213 (1980).

[18] D. M. Leskiw and K. S. Miller "Convergence of Polynomial Least Squares Estimators," in Proceedings of the IEEE, Vol. 70, No. 5, May, 1982.

3

Illustration of the Problem

In this Chapter the problem to be analyzed is illustrated by using a simple example that Julier and Uhlmann employed to motivate their Unscented Kalman Filter (UKF) [1]. A set of unbiased and *independent and identically distributed* (iid) gaussian radar measurements on $P \in \mathbb{E}$ are averaged to estimate (r, a); and the corresponding rectangular pseudo-measurements are averaged to estimate (x, y). The former is optimal in the unbiased and minimum variance sense, but the latter is biased and is noisier. Various linear Kalman filters are then used, and similar problems occur. Finally, the extended Kalman filter is employed: it is less biased and less noisy, but it has some convergence issues. The results are summarized in the concluding Section of this Chapter.

3.1 The Basic Estimation Bias Problem

In the aforementioned "exemplar" of Julier and Uhlmann, the object was located at the point $P \in \mathbb{E}$, having rectangular and radar coordinates $(x, y) = (0, 1)$ and $(r, a) = (1, \pi/2)$. A sequence of mutually independent radar measurements on P was given, $\bar{r}_n = r + \tilde{r}_n$ and $\bar{a}_n = a + \tilde{a}_n$, $n = 1, 2, \cdots, N$, whose errors, $\tilde{r}_n$ and $\tilde{a}_n$, were distributed as $\mathcal{N}(0; \sigma_R^2)$ and $\mathcal{N}(0; \sigma_A^2)$, with $\sigma_R = 0.02$ meters and $\sigma_A = \pi/12$ radians. And the corresponding sequence of rectangular measurements on P was obtained as

$$\overline{x}'_n = \overline{r}_n \cos \overline{a}_n \text{ and } \overline{y}'_n = \overline{r}_n \sin \overline{a}_n . \tag{95}$$

Then, to respectively estimate (r,a) and (x,y), the averages of the two sets of measurements were determined,

$$\hat{\mathbf{r}}_n = \begin{bmatrix} \hat{r}_n \\ \hat{a}_n \end{bmatrix} = \frac{1}{n}\sum_{m=1}^{n}\begin{bmatrix} \overline{r}_m \\ \overline{a}_m \end{bmatrix} \text{ and } \hat{\mathbf{x}}'_n = \begin{bmatrix} \hat{x}'_n \\ \hat{y}'_n \end{bmatrix} = \frac{1}{n}\sum_{m=1}^{n}\begin{bmatrix} \overline{x}'_m \\ \overline{y}'_m \end{bmatrix}. \tag{96}$$

Here, to facilitate comparisons between the two cases, the point P is chosen such that $(x,y)=(r,a)$. Specifically, let $x=r=1$ (meters) and let $y=a=0$ (meters and radians). And, to make the effects of the nonlinearities more pronounced, let $\sigma_A = \pi/6$ (twice the value that Julier and Uhlmann used). In which case the underlying random vector of the radar measurements is distributed according to

$$p_R(\mathbf{r}) = \frac{1}{2\pi(\det \Sigma_R)^{1/2}} \exp\left[-\frac{1}{2}(\mathbf{r}-\boldsymbol{\mu}_R)^T \boldsymbol{\Sigma}_R^{-1}(\mathbf{r}-\boldsymbol{\mu}_R)\right], \tag{97}$$

with

$$\boldsymbol{\mu}_R = \begin{bmatrix} \mu_R \\ \mu_A \end{bmatrix} = \begin{bmatrix} 1 \\ 0 \end{bmatrix} \text{ and } \boldsymbol{\Sigma}_R = \begin{bmatrix} \sigma_R^2 & \sigma_{RA} \\ \sigma_{AR} & \sigma_A^2 \end{bmatrix} = \begin{bmatrix} .004 & 0 \\ 0 & \pi^2/36 \end{bmatrix}. \tag{98}$$

Figure 3-1 illustrates the ensuing radar measurements and rectangular pseudo-measurements for $N = 10{,}000$. Also shown is the sequence of rectangular averages, $(\hat{x}'_n, \hat{y}'_n)$, $n = 1,2,\cdots,N$. The sequence of $(\hat{r},\hat{a})$'s is obscured by the measurements. The two sets of measurements are shown as coordinate-points. The $(\overline{r},\overline{a})$'s are plotted against axes that are linear in range and azimuth; and the $(\overline{x}',\overline{y}')$'s are overlaid using those axes as range and cross-range. (The right-hand side provides an expanded view.) There effects of the nonlinear coordinate transformation are clearly seen: the gaussian set of radar measurements has an elliptical shape; while the corresponding set of rectangular coordinates has a circular shape. Also, the sequence of $(\hat{x}',\hat{y}')$'s appears to be biased – it does not converge to $(1,0)$.

3.1.1 The Scalar-Weight Case

Here the effects of the nonlinear coordinate transformation upon the above estimates are further illustrated. But, in anticipation of the sequel, the *recursive* form of the *weighted* average will now be used. Recall that in the scalar-weight case the coordinate form of a detection is written $(\bar{\mathbf{r}};\mathbf{I}/\bar{w})$ or $(\bar{\mathbf{x}}';\mathbf{I}/\bar{w})$, with $\bar{w}>0$; and the estimates are written $(\hat{\mathbf{r}};\mathbf{I}/\hat{w})$ and $(\hat{\mathbf{x}}';\mathbf{I}/\hat{w})$, with $\hat{w}>0$.

Let a sequence of distinct radar detections be given, $(\bar{\mathbf{r}}_n;\mathbf{I}/\bar{w}_n)$, $n=1,2,\cdots,N$ (the measurements are distinct; the weights may all be the same). Their weighted average is

$$\hat{\mathbf{r}}_N = \frac{1}{\hat{w}_N}\sum_{n=1}^{N}\bar{w}_n\bar{\mathbf{r}}_n \text{ and } \hat{w}_N = \sum_{n=1}^{N}\bar{w}_n . \tag{99}$$

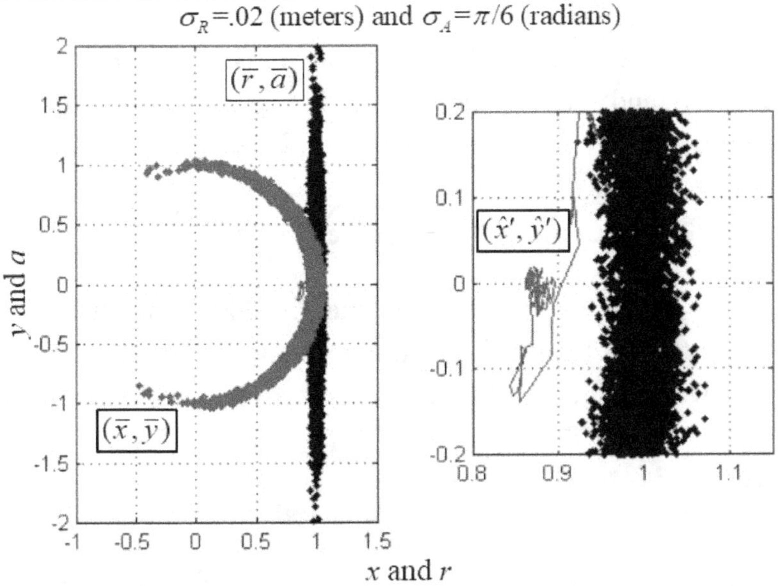

Figure 3-1 The basic estimation bias problem

Note that (96) and (99) become same when all the weights are equal. Indeed,

$$\frac{1}{\hat{w}_n}\sum_{m=1}^{n}\bar{w}_m\bar{\mathbf{r}}_m = \frac{1}{n}\sum_{m=1}^{n}\bar{\mathbf{r}}_m \text{ and } \hat{w}_n = \sum_{m=1}^{n}\bar{w}_m = n\bar{w} . \tag{100}$$

The *recursive fusion* form of (99) is (see the Appendix)

$$\hat{\mathbf{r}}_n = \left(\hat{w}_{n-1}/\hat{w}_n\right)\hat{\mathbf{r}}_{n-1} + \left(\bar{w}_n/\hat{w}_n\right)\bar{\mathbf{r}}_n \text{ and } \hat{w}_n = \hat{w}_{n-1} + \bar{w}_n , \tag{101}$$

$n=1,2,\cdots,N$, with $(\hat{\mathbf{r}}_1;\mathbf{I}/\hat{w}_1)=(\bar{\mathbf{r}}_1;\mathbf{I}/\bar{w}_1)$. And the corresponding *update* form is

$$\hat{\mathbf{r}}_n = \hat{\mathbf{r}}_{n-1} + \left(\bar{w}_n/\hat{w}_n\right)(\bar{\mathbf{r}}_n - \hat{\mathbf{r}}_{n-1}) \text{ and } \hat{w}_n = \hat{w}_{n-1} + \bar{w}_n . \tag{102}$$

These two recursive forms shall be written symbolically as

$$(\hat{\mathbf{r}}_{n-1};\mathbf{I}/\hat{w}_{n-1}) \xrightarrow{(\overline{\mathbf{r}}_n;\mathbf{I}/\overline{w}_n)} (\hat{\mathbf{r}}_n;\mathbf{I}/\hat{w}_n)\,. \tag{103}$$

with $(\hat{\mathbf{r}}_1;\mathbf{I}/\hat{w}_1)=(\overline{\mathbf{r}}_1;\mathbf{I}/\overline{w}_1)$. In the rectangular (pseudo) measurement case the equations have the same form as those given above. And the recursive update is written

$$(\hat{\mathbf{x}}'_{n-1};\mathbf{I}/\hat{w}_{n-1}) \xrightarrow{(\overline{\mathbf{x}}'_n;\mathbf{I}/\overline{w}_n)} (\hat{\mathbf{x}}'_n;\mathbf{I}/\hat{w}_n)\,, \tag{104}$$

with $(\hat{\mathbf{x}}'_1;\mathbf{I}/\hat{w}_1)=(\overline{\mathbf{x}}'_1;\mathbf{I}/\overline{w}_1)$.

To illustrate further the effects of the coordinate transformation upon the estimates, let $\overline{w}_m=\overline{w}$, $n=1,2,\cdots,N$, and partition the $(\overline{r},\overline{a})$'s in Figure 3-1 into M *disjoint* subsets of length L each, with $LM=N$. And then, for $m=1,2,\cdots,M$, determine

$$\hat{\mathbf{r}}_{m,l}=\frac{1}{l}\sum_{k=1}^{l}\overline{\mathbf{r}}_{m,k} \quad\text{and}\quad \hat{\mathbf{x}}'_{m,l}=\frac{1}{l}\sum_{k=1}^{l}\overline{\mathbf{x}}'_{m,k}\,,\ l=1,2,\cdots,L\,. \tag{105}$$

Figure 3-2 shows the case where $M=50$ and $L=200$. Here such sequences shall be called *estimation paths* (of length L). More formally, they are mutually independent Monte Carlo trials, with each trial a Markov chain starting at a given *random draw*.

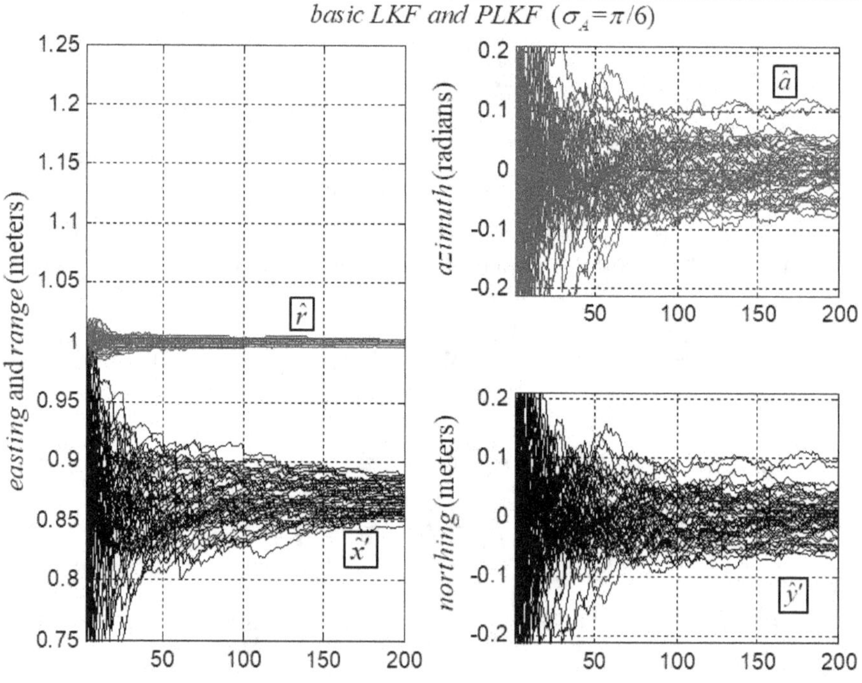

Figure 3-2 The basic estimation paths

Of course, these radar averages may be *converted* into rectangular coordinates to estimate (x, y). That is, after determining each $(\hat{r}_{m,l}, \hat{a}_{m,l})$ in (105), transform them using

$$\hat{x}_{m,l} = \hat{r}_{m,l} \cos \hat{a}_{m,l} \quad \text{and} \quad \hat{y}_{m,l} = \hat{r}_{m,l} \sin \hat{a}_{m,l} \,. \tag{106}$$

Figure 3-3 shows the sequences of these *converted* estimates, denoted by $\hat{x} = \hat{r}\cos\hat{a}$ and $\hat{y} = \hat{r}\sin\hat{a}$, along with the original $(\hat{x}', \hat{y}')$'s of Figure 3-2.

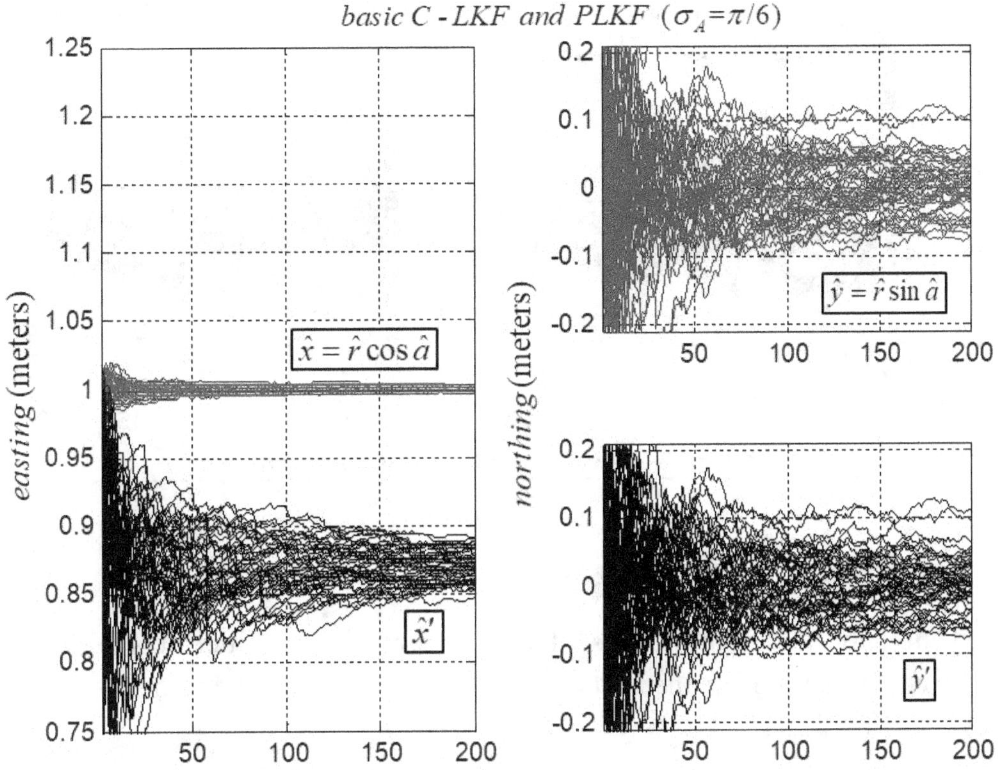

Figure 3-3 The basic rectangular estimation sequences

Similarly, the $(\hat{x}', \hat{y}')$'s may be *converted back* into radar coordinates to provide estimates of (r, a). That is, given $(\hat{\mathbf{x}}'_{m,l})^T = (\hat{x}'_{m,l}, \hat{y}'_{m,l})$ from the second expression of (105), determine

$$\hat{r}'_{m,l} = \sqrt{\hat{x}'^2_{m,l} + \hat{y}'^2_{m,l}} \quad \text{and} \quad \hat{a}'_{m,l} = \arctan\left(\hat{y}'_{m,l}, \hat{x}'_{m,l}\right). \tag{107}$$

Figure 3-4 shows the sequences of these *converted-back* estimates, denoted $\hat{r}' = \sqrt{\hat{x}'^2 + \hat{y}'^2}$ and $\hat{a}' = \arctan(\hat{y}', \hat{x}')$, together with the original $(\hat{r}, \hat{a})$'s of Figure 3-2.

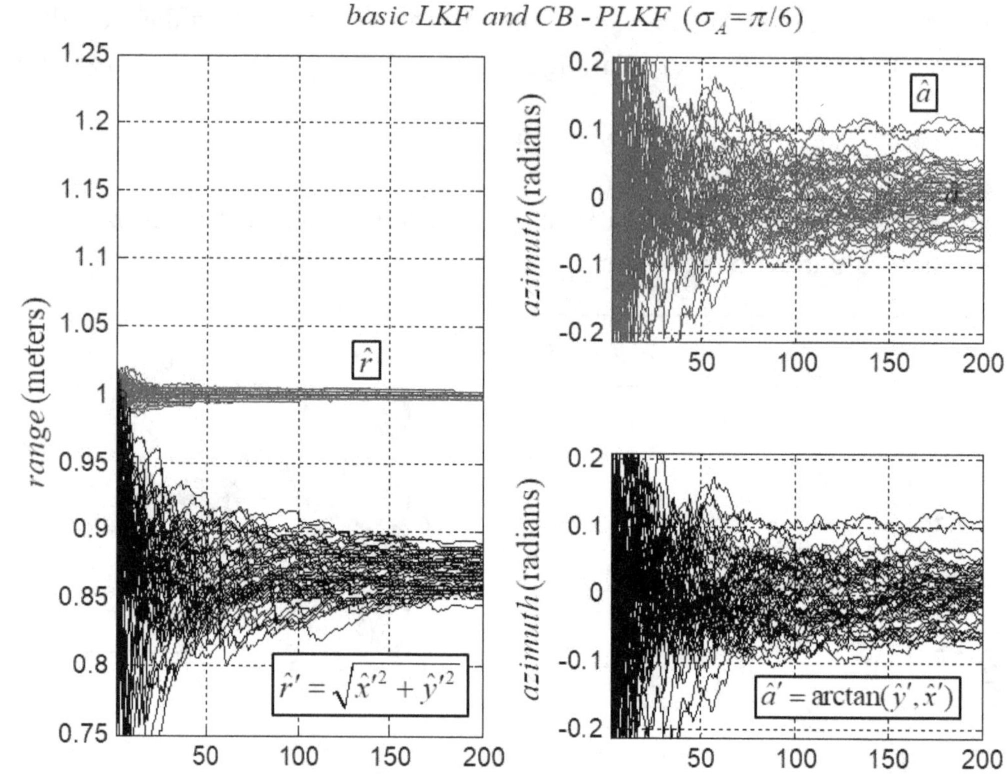

Figure 3-4 The basic radar estimation sequences

3.1.2 The Matrix-Weight Case

In this section the update equations of the Linear Kalman Filter (LKF) are used to determine the estimates. Here the measurements and estimates have associated covariance matrices, and $(\mathbf{h}^{-1};\mathbf{J}^{-1}):(\bar{\mathbf{r}};\bar{\mathbf{R}})\mapsto(\bar{\mathbf{x}}';\bar{\mathbf{X}}')$, where $\bar{\mathbf{R}}=\mathbf{W}^{-1}$ with

$$\mathbf{W}=\begin{bmatrix} w_r & 0 \\ 0 & w_a \end{bmatrix}. \tag{108}$$

Given $(\bar{\mathbf{r}}_n;\bar{\mathbf{R}}_n)$, $n=1,2,\cdots,N$, the batch form of the BLUE is (see the Appendix)

$$\hat{\mathbf{r}}_n=\hat{\mathbf{R}}_n\sum_{k=1}^{n}\bar{\mathbf{R}}_k^{-1}\bar{\mathbf{r}}_k \text{ and } \hat{\mathbf{R}}_n^{-1}=\sum_{k=1}^{n}\bar{\mathbf{R}}_k^{-1}. \tag{109}$$

Its recursive fusion form is

$$\hat{\mathbf{r}}_n=\hat{\mathbf{R}}_n\hat{\mathbf{R}}_{n-1}^{-1}\hat{\mathbf{r}}_{n-1}+\hat{\mathbf{R}}_n\bar{\mathbf{R}}_n^{-1}\bar{\mathbf{r}}_n \text{ and } \hat{\mathbf{R}}_n^{-1}=\hat{\mathbf{R}}_{n-1}^{-1}+\bar{\mathbf{R}}_n^{-1}, \tag{110}$$

and its recursive update form is

$$\hat{\mathbf{r}}_n = \hat{\mathbf{r}}_{n-1} + \hat{\mathbf{R}}_n \bar{\mathbf{R}}_n^{-1}(\bar{\mathbf{r}}_n - \hat{\mathbf{r}}_{n-1}) \text{ and } \hat{\mathbf{R}}_n^{-1} = \hat{\mathbf{R}}_{n-1}^{-1} + \bar{\mathbf{R}}_n^{-1}. \tag{111}$$

These recursive forms are written symbolically as

$$(\hat{\mathbf{r}}_{n-1};\hat{\mathbf{R}}_{n-1}) \xrightarrow{(\bar{\mathbf{r}}_n;\bar{\mathbf{R}}_n)} (\hat{\mathbf{r}}_n;\hat{\mathbf{R}}_n). \tag{112}$$

Similarly, determine

$$\hat{\mathbf{x}}'_n = \hat{\mathbf{X}}'_n \sum_{k=1}^{n} \bar{\mathbf{X}}_k'^{-1} \bar{\mathbf{x}}'_k \text{ and } \hat{\mathbf{X}}_n'^{-1} = \sum_{k=1}^{n} \bar{\mathbf{X}}_k'^{-1}, \tag{113}$$

where $(\mathbf{h}^{-1};\mathbf{J}):(\bar{\mathbf{r}}_n;\bar{\mathbf{R}}_n) \mapsto (\bar{\mathbf{x}}'_n;\bar{\mathbf{X}}'_n)$, $n = 1, 2, \cdots, N$. That is,

$$\bar{\mathbf{x}}'_n = \mathbf{h}^{-1}(\bar{\mathbf{r}}_n) \text{ and } \bar{\mathbf{X}}'_n = \mathbf{J}^{-1}(\bar{\mathbf{r}}_n)\bar{\mathbf{R}}_n \mathbf{J}^{-T}(\bar{\mathbf{r}}_n). \tag{114}$$

Expressions similar to those in (110) and (111) follow. Also,

$$(\hat{\mathbf{x}}'_{n-1};\hat{\mathbf{X}}'_{n-1}) \xrightarrow{(\bar{\mathbf{x}}'_n;\bar{\mathbf{X}}'_n)} (\hat{\mathbf{x}}'_n;\hat{\mathbf{X}}'_n) \tag{115}$$

or, equivalently,

$$(\hat{\mathbf{x}}'_{n-1};\hat{\mathbf{X}}'_{n-1}) \xrightarrow{(\mathbf{h}^{-1};\mathbf{J}):(\bar{\mathbf{r}}_n;\bar{\mathbf{R}}_n)\mapsto(\bar{\mathbf{x}}'_n;\bar{\mathbf{X}}'_n)} (\hat{\mathbf{x}}'_n;\hat{\mathbf{X}}'_n). \tag{116}$$

Note that, when $\bar{\mathbf{R}}_n = \mathbf{\Sigma}_{\mathbf{R}}$, $n = 1, 2, \cdots, N$, see (98), the matrix-weight estimates determined by (109) are the same as those determined in the scalar-weight case,

$$\hat{\mathbf{r}}_N = \frac{1}{N}\sum_{n=1}^{N} \mathbf{\Sigma}_{\mathbf{R}}^{-1}\bar{\mathbf{r}}_n = \frac{1}{N}\sum_{n=1}^{N} \bar{\mathbf{r}}_n \text{ and } \hat{\mathbf{R}}_N = \mathbf{\Sigma}_{\mathbf{R}}/N. \tag{117}$$

But for $n = 2, 3, \cdots, N$, the rectangular estimates determined by (113) are not the same as the ones in the corresponding scalar-weight case: the $\bar{\mathbf{r}}$'s in (114) are all distinct, and so the $\bar{\mathbf{X}}'$'s are also distinct.

Figure 3-5 illustrates the estimates determined by (113) for the equi-weighted case, $\bar{\mathbf{R}}_n = \mathbf{\Sigma}_{\mathbf{R}}$, $n = 1, 2, \cdots, N$, denoted by $\hat{x}'^{(m)}$ and $\hat{y}'^{(m)}$ – the superscript "(m)" indicates the *matrix-weight* case. Here the same subsets of measurements used in Figure 3-2 are being reused, and so the estimates shown there are repeated, but now they are labeled $\hat{x}'^{(s)}$ and $\hat{y}'^{(s)}$ – the superscript "(s)" indicates the *scalar-weight* case. Also shown in Figure 3-5 are the corresponding estimates of (x, y) determined by (106). Figure 3-6 provides the estimates in Figure 3-5 converted back into radar coordinates.

Note that, as in the scalar-weight case, the matrix-weight estimation equations, (112) and (115), have the same form – both are LKF's. But in the sequel, just the one that uses the radar measurements directly, (112), shall be called an LKF. The other case, (115), which uses the rectangular pseudo-measurements, shall be called a Pseudo-LKF (PLKF). Also, the case where the LKF estimates of (r,a) are *converted* into rectangular coordinates shall be called a Converted LKF (C-LKF); and the case where PLKF estimates of (x,y) are *converted back* into radar coordinates shall be called a Converted Back PLKF (CB-PLKF).

3.2 The Extended Kalman Filter

The LKF and PLKF defined above determine their estimates in the same coordinate system as the measurements they were given. That is,

$$(\hat{\mathbf{r}}_{n-1};\hat{\mathbf{R}}_{n-1})\xrightarrow{(\bar{\mathbf{r}}_n;\bar{\mathbf{R}}_n)}(\hat{\mathbf{r}}_n;\hat{\mathbf{R}}_n) \text{ and } (\hat{\mathbf{x}}'_{n-1};\hat{\mathbf{X}}'_{n-1})\xrightarrow{(\bar{\mathbf{x}}'_n;\bar{\mathbf{X}}'_n)}(\hat{\mathbf{x}}'_n;\hat{\mathbf{X}}'_n).$$

In contrast, the Extended Kalman Filter (EKF) uses the radar measurements directly to estimate (x,y). Symbolically, such is written

$$(\hat{\mathbf{x}}''_{n-1};\hat{\mathbf{X}}''_{n-1})\xrightarrow{(\bar{\mathbf{r}}_n;\bar{\mathbf{R}}_n)}(\hat{\mathbf{x}}''_n;\hat{\mathbf{X}}''_n). \tag{118}$$

In particular, given a prior track in rectangular coordinates, $(\hat{\mathbf{x}}''_{n-1};\hat{\mathbf{X}}''_{n-1})$, and given a subsequent detection in radar coordinates, $(\bar{\mathbf{r}}_n;\bar{\mathbf{R}}_n)$, the EKF update is [7]

$$\hat{\mathbf{x}}''_n=\hat{\mathbf{x}}''_{n-1}+\mathbf{K}_{n-1}(\bar{\mathbf{r}}_n-\mathbf{h}(\hat{\mathbf{x}}''_{n-1})) \text{ and } \hat{\mathbf{X}}''_n=\left[\mathbf{I}-\mathbf{K}_{n-1}\mathbf{J}(\hat{\mathbf{x}}_{n-1})\right]\hat{\mathbf{X}}''_{n-1}, \tag{119}$$

$$\mathbf{K}_n\equiv\hat{\mathbf{X}}''_{n-1}\mathbf{J}^T(\hat{\mathbf{x}}_{n-1})\left[\mathbf{J}(\hat{\mathbf{x}}_{n-1})\hat{\mathbf{X}}''_{n-1}\mathbf{J}^T(\hat{\mathbf{x}}_{n-1})+\bar{\mathbf{R}}\right]^{-1}. \tag{120}$$

To begin this recursion, let $(\hat{\mathbf{x}}''_1;\hat{\mathbf{X}}''_1)\equiv(\bar{\mathbf{x}}'_1;\bar{\mathbf{X}}'_1)$.

Figure 3-7 illustrates the EKF estimates, along with the corresponding scalar- and matrix-weight PLKF cases that were shown earlier – the same sets of detections used above are being reused here. Figure 3-8 shows the corresponding *converted back* case, CB-EKF, together with the CB-PLKF.

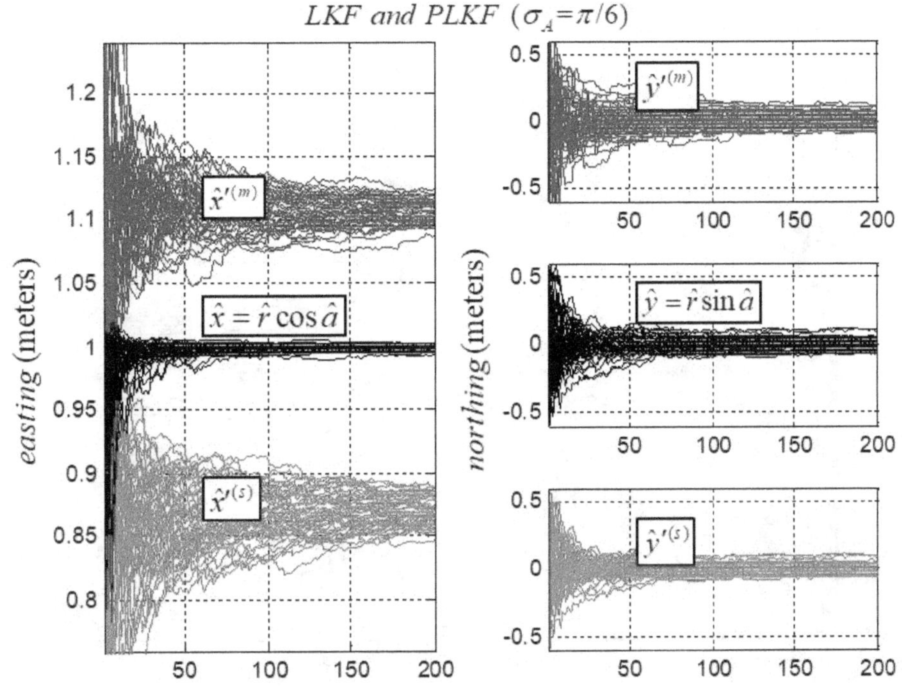

Figure 3-5 The scalar- and matrix-weight cases (rectangular coordinates)

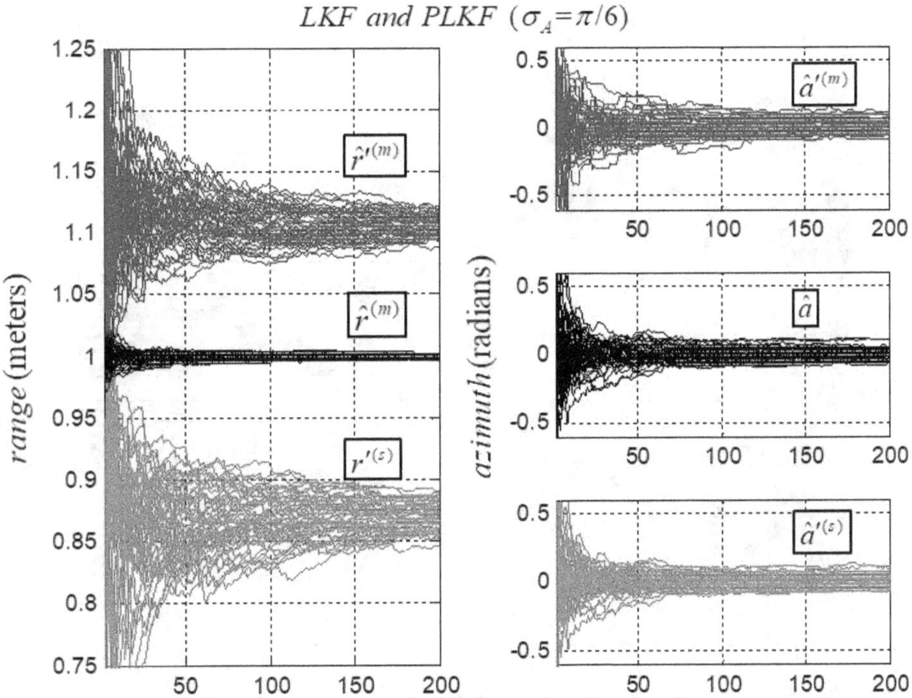

Figure 3-6 The scalar- and matrix-weight cases (radar coordinates)

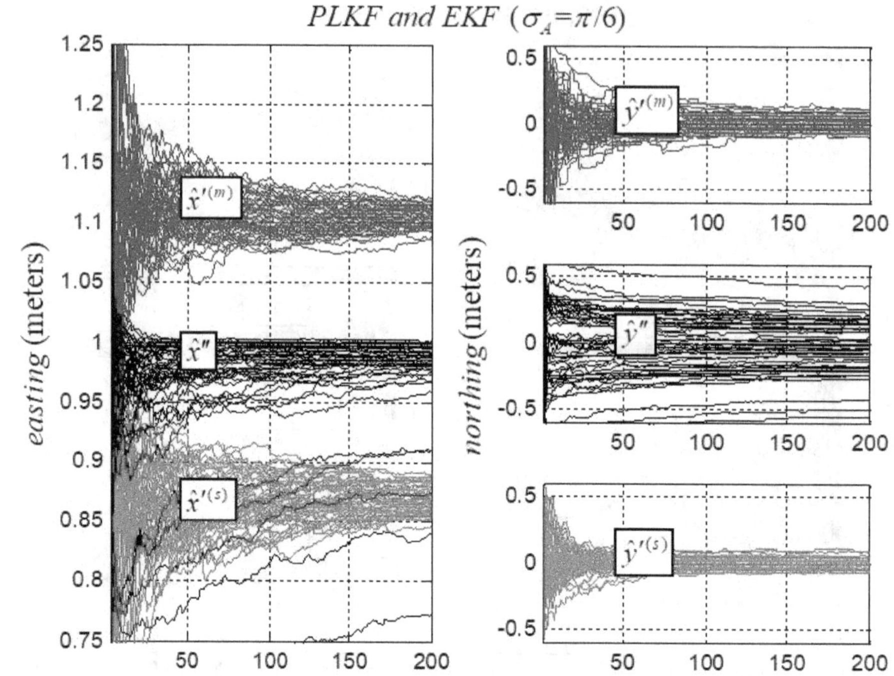

Figure 3-7 PLKF and EKF estimates of (x, y)

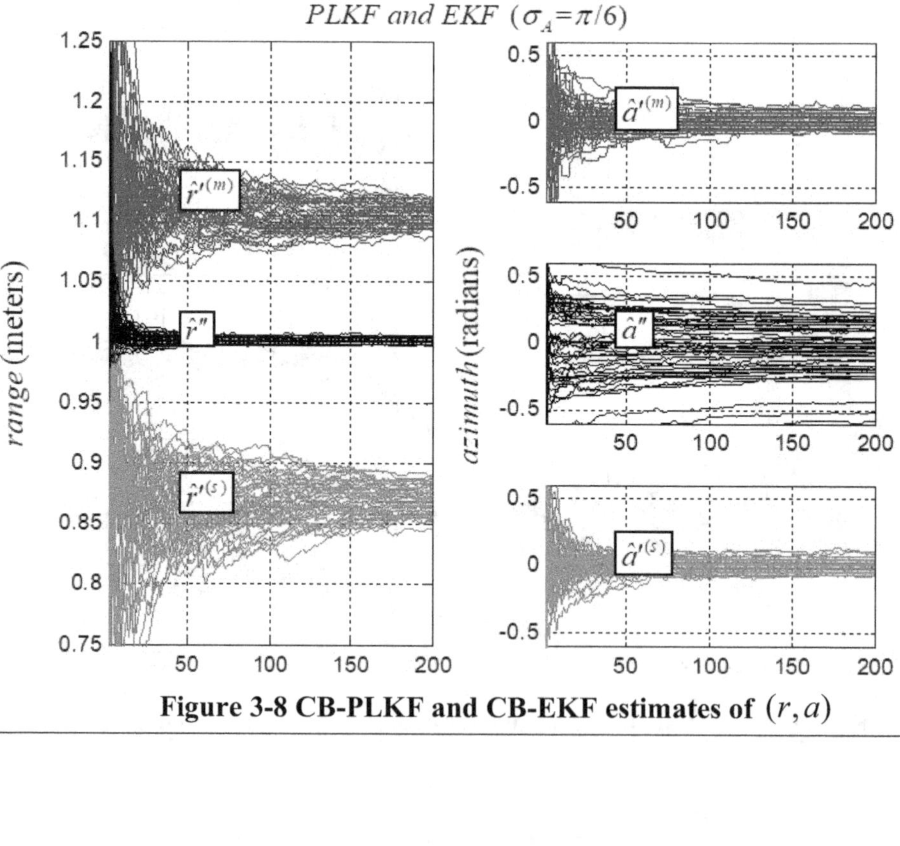

Figure 3-8 CB-PLKF and CB-EKF estimates of (r, a)

3.3 Discussion of the Basic Estimation Problem

The above results are summarized as follows. The LKF and C-LKF estimates seem to be unbiased and have the least variance. The PLKF and CB-PLKF estimates seem to be biased and noisier than the corresponding LKF and C-LKF ones. And the biases of the scalar- and matrix-weight PLKF estimates have the opposite sense) The EKF seems to be less biased and less noisy than the PLKF, but it appears to have convergence issues.

Now $\sigma_A = \pi/6$ in the cases shown above, but Julier and Uhlmann used $\sigma_A = \pi/12$. And so repeat the above cases using their smaller σ_A. Figure 3-9 provides those PLKF and EKF estimates; and Figure 3-10 provides the corresponding CB-PLKF and CB-EKF cases. Note that the EKF now appears to be unbiased, and the PLKF seems to be less biased. Also, the EKF convergence problem has been mitigated somewhat.

Finally, Figure 3-11 and Figure 3-12 respectively compare the sample means for the rectangular and radar coordinate sets of cases that have been shown in this chapter. And a comparison of the sample standard deviations is given below in Figure 3-13 and Figure 3-14. But, for the sake of comparison, they are normalized by $\sigma_X = \sigma_R = .02$ and by $\sigma_Y = \sigma_A$, with either $\sigma_A = \pi/6$ or $\sigma_A = \pi/12$ (respectively the solid and dashed curves). In these figures the solid curves denote the $\sigma_A = \pi/6$ sub-cases, and the dashed ones denote the $\sigma_A = \pi/12$ sub-cases. The (normalized) optimal case is also shown, the smooth curves, $1/\sqrt{n}$. Also, the component forms of the underlying random vectors for $(\hat{x}_l', \hat{y}_l')$ and $(\hat{r}_l', \hat{a}_l')$ are respectively

$$\hat{\boldsymbol{X}}'^T(l) = \left(\hat{X}'(l),\ \hat{Y}'(l)\right) \quad \text{and} \quad \hat{\boldsymbol{R}}'^T(l) = \left(\hat{R}'(l),\ \hat{A}'(l)\right) \tag{121}$$

– with those for the other estimates are written similarly.

For later reference, here the sample means and sample covariance matrices for the LKF estimates are respectively determined as

$$\hat{\boldsymbol{\mu}}_{\hat{\boldsymbol{R}}(l)} = \begin{bmatrix} \hat{\mu}_{\hat{R}(l)} \\ \hat{\mu}_{\hat{A}(l)} \end{bmatrix} = \frac{1}{M}\sum_{m=1}^{M}\begin{bmatrix} \hat{r}_{m,l} \\ \hat{a}_{m,l} \end{bmatrix} = \frac{1}{M}\sum_{m=1}^{M}\hat{\mathbf{r}}_{m,l} \tag{122}$$

and

$$\boldsymbol{\Sigma}_{\hat{\boldsymbol{R}}(l)} = \begin{bmatrix} \hat{\sigma}^2_{\hat{R}(l)} & \hat{\sigma}_{\hat{R}(l)\hat{A}(l)} \\ \hat{\sigma}_{\hat{R}(l)\hat{A}(l)} & \hat{\sigma}^2_{\hat{A}(l)} \end{bmatrix} = \frac{1}{M-1}\sum_{m=1}^{M}\left[\hat{\mathbf{r}}_{m,l} - \hat{\boldsymbol{\mu}}_{\hat{\boldsymbol{R}}(l)}\right]\left[\hat{\mathbf{r}}_{m,l} - \hat{\boldsymbol{\mu}}_{\hat{\boldsymbol{R}}(l)}\right]^T. \tag{123}$$

– with those for the other estimates are written similarly.

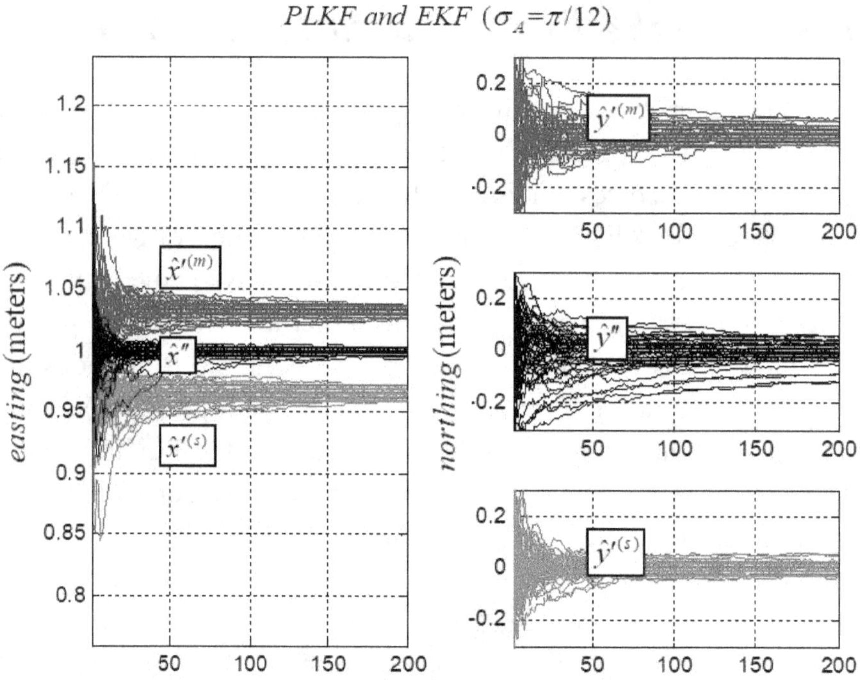

Figure 3-9 PLKF and EKF rectangular estimates ($\sigma_A = \pi/12$)

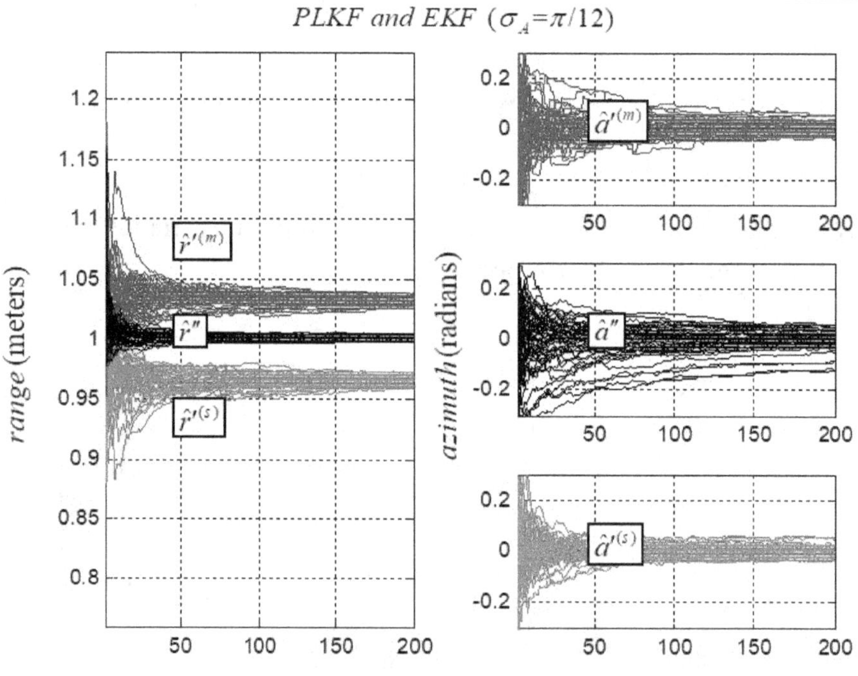

Figure 3-10 CB-PLKF and CB-EKF radar estimates ($\sigma_A = \pi/12$)

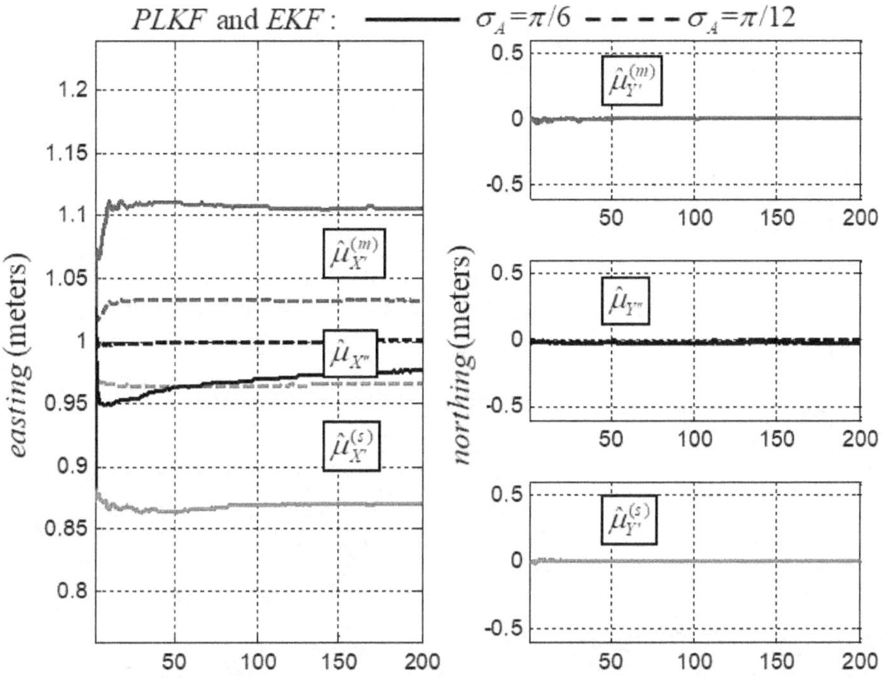

Figure 3-11 Comparison of the sample means (rectangular cases)

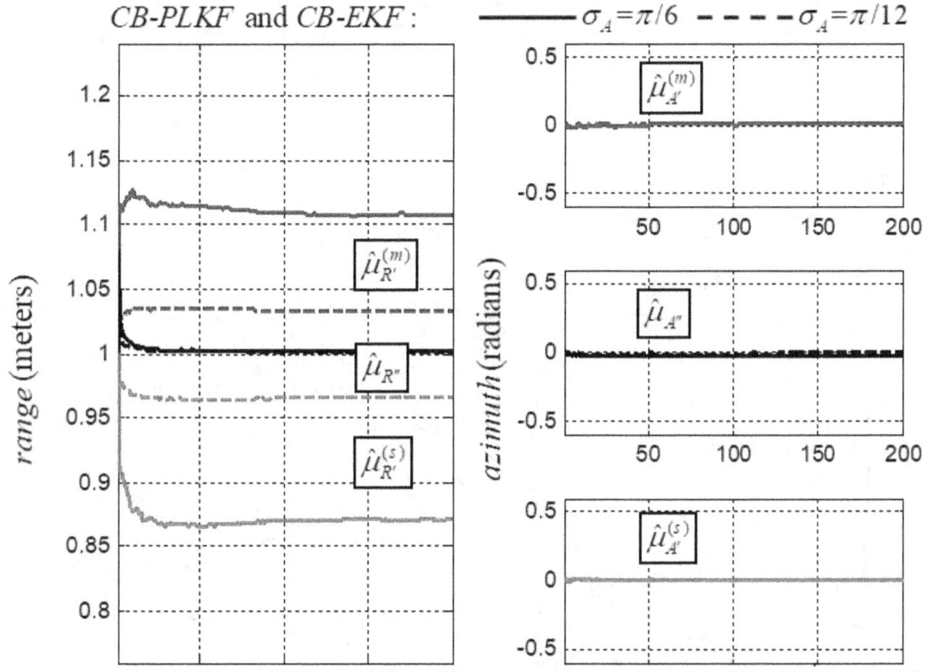

Figure 3-12 Comparison of the sample means (radar cases)

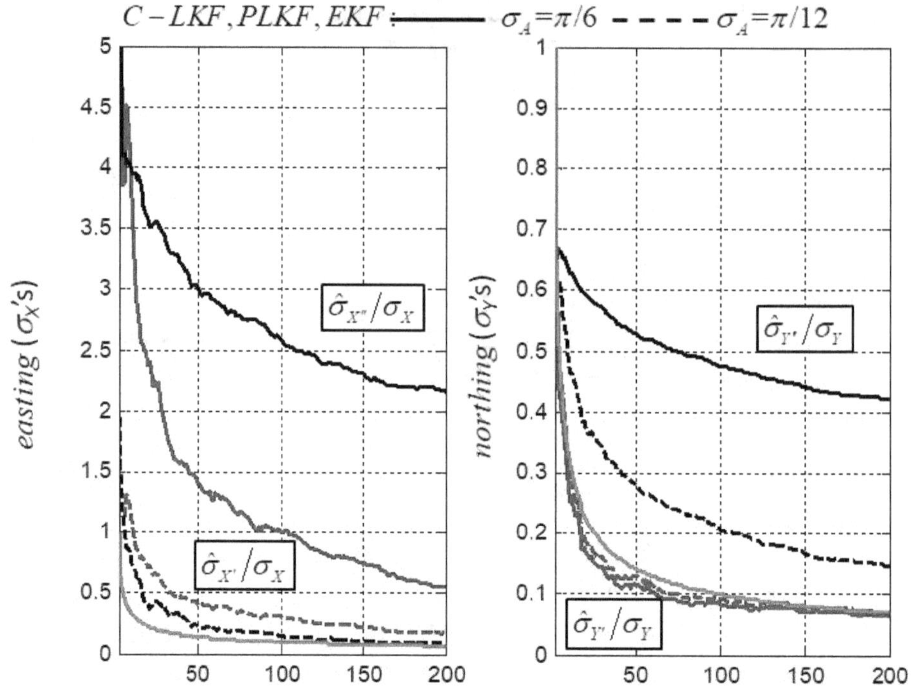

Figure 3-13 Normalized sample standard deviations (rectangular cases)

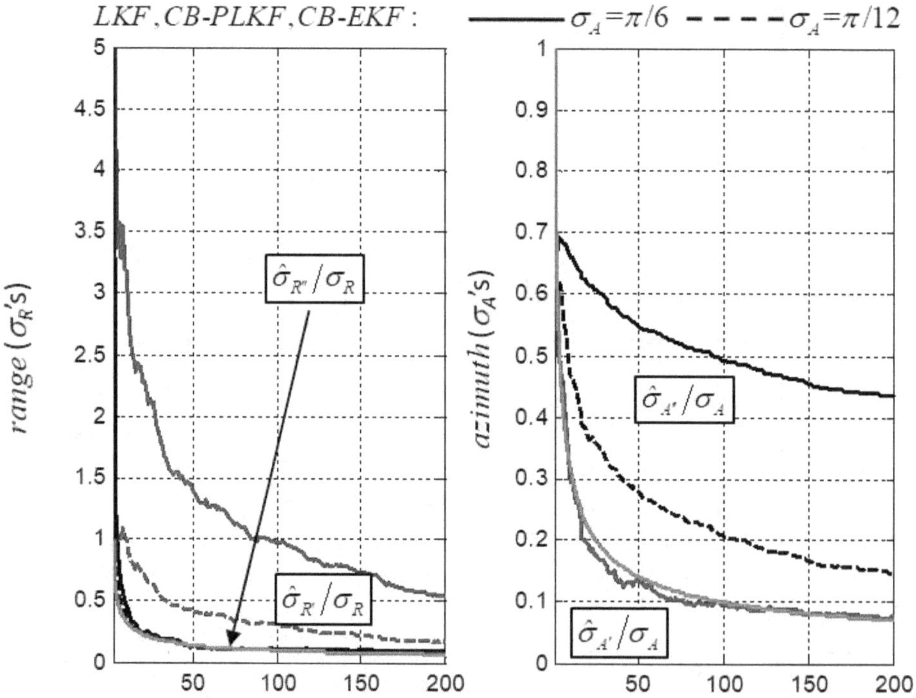

Figure 3-14 Normalized sample standard deviations (radar cases)

3.4 *Chapter 3 References*

[1] S. Julier and J. Uhlmann, "A New Extension of the Kalman Filter to Nonlinear Systems," in Proceedings of AeroSense: Symposium Aerospace/Defense Sensing, Simulation and Control, Vol. 3373, pp. 183-193 (1997).

4

Analysis of the PLKF Update

In the previous Chapter the PLKF was defined to be an LKF that uses rectangular pseudo-measurements to estimate (x, y). And there two PLKF cases were distinguished: scalar-weight and matrix-weight. Both PLKF's were seen to be biased, and usually worse than simply converting the (optimal) LKF estimates of (r, a) using $\hat{x} = \hat{r}\cos\hat{a}$ and $\hat{y} = \hat{r}\sin\hat{a}$ – the C-LKF case. Here those estimators are discussed more formally, and certain analytic results are obtained for use in the sequel.

4.1 The Pseudo-Measurement Errors

Recall that the canonical measurement model used in radar tracking is $\bar{\mathbf{r}} = \mathbf{r} + \tilde{\mathbf{r}}$, where $\mathbf{r} = \mathbf{h}(\mathbf{x})$, with the measurement error, $\tilde{\mathbf{r}}$, tacitly assumed to be independent of $\mathbf{x}$. Similarly, the model of the rectangular pseudo-measurements is $\bar{\mathbf{x}}' \equiv \mathbf{x} + \tilde{\mathbf{x}}'$. But, unlike the radar measurements, the components of $\tilde{\mathbf{x}}'$ are not independent of $\mathbf{x}$. Indeed, substituting $r + \tilde{r}$ and $a + \tilde{a}$ into $\bar{x}' = \bar{r}\cos\bar{a}$ and $\bar{y}' = \bar{r}\sin\bar{a}$ yields

$$\bar{x}' = r\cos a\cos\tilde{a} - r\sin a\sin\tilde{a} + \frac{\tilde{r}}{r}(r\cos a\cos\tilde{a} - r\sin a\sin\tilde{a})$$

and

$$\overline{y}' = r\sin a\cos\tilde{a} + r\cos a\sin\tilde{a} + \frac{\tilde{r}}{r}(r\sin a\cos\tilde{a} + r\cos a\sin\tilde{a}) \, ,$$

which simplify to

$$\begin{bmatrix} \overline{x}' \\ \overline{y}' \end{bmatrix} = (1+\tilde{r}/r)\begin{bmatrix} \cos\tilde{a} & -\sin\tilde{a} \\ +\sin\tilde{a} & \cos\tilde{a} \end{bmatrix}\begin{bmatrix} x \\ y \end{bmatrix}. \tag{124}$$

That is, the errors of the pseudo-measurements depend nonlinearly upon both $\mathbf{x}$ and $\tilde{\mathbf{r}}$,

$$\begin{bmatrix} \tilde{x}' \\ \tilde{y}' \end{bmatrix} \equiv \begin{bmatrix} \overline{x}' - x \\ \overline{y}' - y \end{bmatrix} = \begin{bmatrix} \cos\tilde{a} - 1 & -\sin\tilde{a} \\ +\sin\tilde{a} & \cos\tilde{a} - 1 \end{bmatrix}\begin{bmatrix} x \\ y \end{bmatrix} + \tilde{r}\begin{bmatrix} \cos\overline{a} \\ \sin\overline{a} \end{bmatrix}. \tag{125}$$

4.1.1 The Expected Value of the Pseudo-Measurements

The underlying random vectors of $\overline{\mathbf{r}} = \mathbf{r} + \tilde{\mathbf{r}}$ are related as $\boldsymbol{R} = \mathbf{r} + \tilde{\boldsymbol{R}}$. Let $\tilde{\boldsymbol{R}}^T = (\tilde{R}, \tilde{A})$, with $\tilde{R}$ and $\tilde{A}$ independent and distributed as $\tilde{R} \sim \mathcal{N}(0;\sigma_R^2)$ and $\tilde{A} \sim \mathcal{N}(0;\sigma_A^2)$. The underlying random vector of the rectangular pseudo-measurements is $\boldsymbol{X}' = \mathbf{h}^{-1}(\boldsymbol{R})$, and its components are

$$X' = R\cos A \text{ and } Y' = R\sin A. \tag{126}$$

But since R and A are independent, the expected values of these expressions factor as

$$\mu_{X'} \equiv \mathcal{E}X' = (\mathcal{E}R)(\mathcal{E}\cos A) \text{ and } \mu_{Y'} \equiv \mathcal{E}Y' = (\mathcal{E}R)(\mathcal{E}\sin A). \tag{127}$$

And, since $\mathcal{E}R = r$ and $A = a + \tilde{A}$,

$$\begin{bmatrix} \mu_{X'} \\ \mu_{Y'} \end{bmatrix} = r\begin{bmatrix} \cos a & -\sin a \\ \sin a & \cos a \end{bmatrix}\begin{bmatrix} \mathcal{E}\cos\tilde{A} \\ \mathcal{E}\sin\tilde{A} \end{bmatrix}. \tag{128}$$

By symmetry, $\mathcal{E}\sin\tilde{A} = 0$. Thus, $\mathcal{E}\mathbf{X}' = \mathbf{x}\mathcal{E}\cos\tilde{A}$, where

$$\mathcal{E}\cos\tilde{A} = \int_{-\infty}^{+\infty}\cos(\alpha)p_{\tilde{A}}(\alpha)d\alpha \, .$$

This integral can be evaluated using a technique outlined in [1]. Let

$$I(t) = \int_{-\infty}^{+\infty} [\cos(\alpha t)] p_{\tilde{A}}(\alpha) d\alpha \,. \tag{129}$$

Since $\cos(\alpha t)$ is continuous at any finite α and t,

$$\frac{d}{dt} I(t) = \int_{-\infty}^{+\infty} \left[\frac{\partial}{\partial t} \cos(\alpha t) \right] p_{\tilde{A}}(\alpha) d\alpha = -\int_{-\infty}^{+\infty} \alpha \sin(\alpha t) p_{\tilde{A}}(\alpha) d\alpha \,.$$

Integration by parts, $\int u dv = uv - \int v du$, where $u = \sin(\alpha t)$ and $v = \sigma_{\tilde{A}}^2 p_{\tilde{A}}(a)$, yields

$$\frac{d}{dt} I(t) = \sin(\alpha t) \sigma_{\tilde{A}}^2 p_{\tilde{A}}(\alpha) \Big|_{\alpha=-\infty}^{+\infty} - t\sigma_{\tilde{A}}^2 \int_{-\infty}^{+\infty} \cos(\alpha t) p_{\tilde{A}}(\alpha) d\alpha \,. \tag{130}$$

In the above expression, since $p_{\tilde{A}}$ is of order $e^{-\alpha^2}$, the first summand is zero,

$$\lim_{a \to \infty} \sin(\alpha t) \sigma_{\tilde{A}}^2 p_{\tilde{A}}(\alpha) \Big|_{\alpha=-a}^{\alpha=+a} = 0 \,.$$

And the second summand is

$$t\sigma_{\tilde{A}}^2 \int_{-\infty}^{+\infty} \cos(\alpha t) p_{\tilde{A}}(\alpha) d\alpha = t\sigma_{\tilde{A}}^2 I(t) \,.$$

Thus,

$$\frac{d}{dt} I(t) = -t\sigma_{\tilde{A}}^2 I(t) \,.$$

The solution to this differential equation is

$$I(t) = I_0 \exp(-t^2 \, \sigma_{\tilde{A}}^2 / 2) \,, \tag{131}$$

where I_0 is an integration constant that is independent of t. Let $t = 0$ in (129) to obtain $I_0 = 1$; and then let $t = 1$ with $I_0 = 1$ to obtain

$$\mathcal{E} \cos \tilde{A} = \int_{-\infty}^{+\infty} (\cos \alpha) p_{\tilde{A}}(\alpha) d\alpha = e^{-\sigma_{\tilde{A}}^2/2} \,. \tag{132}$$

Thus, the expected value of $\boldsymbol{X}'$ is

$$\boldsymbol{\mu}_{X'} = \begin{bmatrix} \mu_{X'} \\ \mu_{Y'} \end{bmatrix} = r \begin{bmatrix} \cos a \\ \sin a \end{bmatrix} e^{-\sigma_A^2/2} = \begin{bmatrix} x \\ y \end{bmatrix} e^{-\sigma_A^2/2}. \tag{133}$$

Note that $\boldsymbol{\mu}_{X'} = \mathbf{x}e^{-\sigma_A^2/2} \neq \mathbf{x}$ when $\sigma_A \neq 0$.

4.1.2 The Covariance Matrix of the Pseudo-Measurements

By definition, the covariance matrix of the rectangular pseudo-measurements is

$$\boldsymbol{\Sigma}_{X'} \equiv \mathcal{E}(\boldsymbol{X}' - \boldsymbol{\mu}_{X'})(\boldsymbol{X}' - \boldsymbol{\mu}_{X'})^T = \begin{bmatrix} \sigma_{X'}^2 & \sigma_{X'Y'} \\ \sigma_{X'Y'} & \sigma_{Y'}^2 \end{bmatrix}. \tag{134}$$

Which is equivalent to

$$\mathcal{E}(\boldsymbol{X}' - \boldsymbol{\mu}_{X'})(\boldsymbol{X}' - \boldsymbol{\mu}_{X'})^T = \mathcal{E}\boldsymbol{X}'\boldsymbol{X}'^T - \boldsymbol{\mu}_{X'}\boldsymbol{\mu}_{X'}^T. \tag{135}$$

From (133),

$$\boldsymbol{\mu}_{X'}\boldsymbol{\mu}_{X'}^T = \begin{bmatrix} \mu_{X'} & \mu_{X'}\mu_{Y'} \\ \mu_{Y'}\mu_{X'} & \mu_{Y'} \end{bmatrix} = e^{-\sigma_A^2} \begin{bmatrix} x^2 & xy \\ yx & y^2 \end{bmatrix}. \tag{136}$$

And so, to evaluate $\boldsymbol{\Sigma}_{X'}$ just $\mathcal{E}\boldsymbol{X}'\boldsymbol{X}'^T$ remains to be determined.

Using (127) and the mutual independence of R and A,

$$\mathcal{E}\boldsymbol{X}'\boldsymbol{X}'^T = \mathcal{E}\begin{bmatrix} X'^2 & X'Y' \\ Y'X' & Y'^2 \end{bmatrix} = (\mathcal{E}R^2)\mathcal{E}\begin{bmatrix} \cos^2 A & \cos A \sin A \\ \cos A \sin A & \sin^2 A \end{bmatrix}. \tag{137}$$

Of course,

$$2\begin{bmatrix} \cos^2 A & \cos A \sin A \\ \cos A \sin A & \sin^2 A \end{bmatrix} = \begin{bmatrix} 1 & 0 \\ 0 & 1 \end{bmatrix} + \begin{bmatrix} +\cos 2A & \sin 2A \\ \sin 2A & -\cos 2A \end{bmatrix}, \tag{138}$$

and

$$\begin{bmatrix} +\cos 2A & \sin 2A \\ \sin 2A & -\cos 2A \end{bmatrix} = \cos 2\tilde{A} \begin{bmatrix} +\cos 2a & \sin 2a \\ \sin 2a & -\cos 2a \end{bmatrix} + \sin 2\tilde{A} \begin{bmatrix} -\sin 2a & \cos 2a \\ \cos 2a & \sin 2a \end{bmatrix}. \tag{139}$$

By symmetry $\mathcal{E}\sin 2\tilde{A}=0$. And using (132) with $\beta=2\alpha$, $\tilde{B}=2\tilde{A}$, $\sigma_{\tilde{B}}^2=4\sigma_{\tilde{A}}^2$,

$$\mathcal{E}\cos 2\tilde{A}=\int_{-\infty}^{+\infty}(\cos 2\alpha)p_{\tilde{A}}(\alpha)d\alpha=\int_{-\infty}^{+\infty}(\cos\beta)\frac{2\exp(-\beta^2/2\sigma_{\tilde{B}}^2)}{\sigma_B\sqrt{2\pi}}d\beta/2=e^{-\sigma_{\tilde{B}}^2/2}.$$

Thus, since $\sigma_{\tilde{B}}^2/2=2\sigma_{\tilde{A}}^2$, the expected value of (139) is

$$\mathcal{E}\begin{bmatrix}+\cos 2A & \sin 2A\\ \sin 2A & -\cos 2A\end{bmatrix}=e^{-2\sigma_{\tilde{A}}^2}\begin{bmatrix}+\cos 2a & \sin 2a\\ \sin 2a & -\cos 2a\end{bmatrix}. \tag{140}$$

Substituting the above result into the expected value of (138) yields

$$\mathcal{E}\begin{bmatrix}\cos^2 A & \cos A\sin A\\ \cos A\sin A & \sin^2 A\end{bmatrix}=\frac{1}{2}\begin{bmatrix}1 & 0\\ 0 & 1\end{bmatrix}+\frac{e^{-2\sigma_{\tilde{A}}^2}}{2}\begin{bmatrix}+\cos 2a & \sin 2a\\ \sin 2a & -\cos 2a\end{bmatrix}. \tag{141}$$

And using this result in (141), along with $x/r=\cos a$ and $y/r=\sin a$, and the double angle identities for the cosine and sine functions, gives

$$\mathcal{E}\begin{bmatrix}\cos^2 A & \cos A\sin A\\ \cos A\sin A & \sin^2 A\end{bmatrix}=\frac{1-e^{-2\sigma_{\tilde{A}}^2}}{2}\begin{bmatrix}1 & 0\\ 0 & 1\end{bmatrix}+\frac{e^{-2\sigma_{\tilde{A}}^2}}{r^2}\begin{bmatrix}x^2 & xy\\ xy & y^2\end{bmatrix}. \tag{142}$$

Also, $\sigma_{\tilde{R}}^2=\mathcal{E}R^2-r^2$. And so the matrix defined by (137) is

$$\mathcal{E}\mathbf{X}'\mathbf{X}'^T=(\sigma_{\tilde{R}}^2+r^2)\left(\frac{1-e^{-2\sigma_{\tilde{A}}^2}}{2}\begin{bmatrix}1 & 0\\ 0 & 1\end{bmatrix}+\frac{e^{-2\sigma_{\tilde{A}}^2}}{r^2}\begin{bmatrix}x^2 & xy\\ xy & y^2\end{bmatrix}\right). \tag{143}$$

Therefore, using (136) and (143) in (135), the covariance matrix of $\boldsymbol{X}'$ is

$$\boldsymbol{\Sigma}_{X'}=\left(\sigma_{\tilde{R}}^2+r^2\right)\left\{\frac{1-e^{-2\sigma_{\tilde{A}}^2}}{2}\begin{bmatrix}1 & 0\\ 0 & 1\end{bmatrix}+\frac{e^{-2\sigma_{\tilde{A}}^2}}{r^2}\begin{bmatrix}x^2 & xy\\ xy & y^2\end{bmatrix}\right\}-e^{-\sigma_{\tilde{A}}^2}\begin{bmatrix}x^2 & xy\\ yx & y^2\end{bmatrix}. \tag{144}$$

Note that $\boldsymbol{\Sigma}_{X'}$ is a function of $\mathbf{x}$. And so, more explicitly, write

$$\boldsymbol{\Sigma}_{X'}(\mathbf{x})=\begin{bmatrix}(\Sigma_{X'})_{xx}(x,y) & (\Sigma_{X'})_{xy}(x,y)\\ (\Sigma_{X'})_{yx}(x,y) & (\Sigma_{X'})_{yy}(x,y)\end{bmatrix}. \tag{145}$$

Alternatively, using $\mathbf{x}=\mathbf{h}^{-1}(\mathbf{r})$, one may write

$$\mathbf{\Sigma}_{X'}(\mathbf{r}) = \begin{bmatrix} (\Sigma_{X'})_{xx}(r,a) & (\Sigma_{X'})_{xy}(r,a) \\ (\Sigma_{X'})_{yx}(r,a) & (\Sigma_{X'})_{yy}(r,a) \end{bmatrix}. \tag{146}$$

In (145) and (146) the subscripts on the $(\mathbf{\Sigma}_{X'})$'s index the components of the matrix; and the arguments serve to specify both the parameterization and the point at which the matrices are being evaluated.

Both forms are needed for the sequel. And so, given (144), the component form of (146) is determined as follows. First, rewrite (143) as

$$\frac{r^2\mathcal{E}\mathbf{X}'\mathbf{X}'^T}{e^{\sigma_A^2}(\sigma_R^2+r^2)} = \frac{e^{\sigma_A^2}}{2}\begin{bmatrix} x^2+y^2 & 0 \\ 0 & x^2+y^2 \end{bmatrix} + \frac{e^{-\sigma_A^2}}{2}\begin{bmatrix} 2x^2 & 2xy \\ 2xy & 2y^2 \end{bmatrix} - \frac{e^{-\sigma_A^2}}{2}\begin{bmatrix} x^2+y^2 & 0 \\ 0 & x^2+y^2 \end{bmatrix}.$$

Then use $\cosh\sigma = (e^{\sigma}+e^{-\sigma})/2$ and $\sinh\sigma = (e^{\sigma}-e^{-\sigma})/2$ to obtain

$$\frac{r^2\mathcal{E}\mathbf{X}'\mathbf{X}'^T}{e^{\sigma_A^2}(\sigma_R^2+r^2)} = x^2\begin{bmatrix} \cosh\sigma_A^2 & 0 \\ 0 & \sinh\sigma_A^2 \end{bmatrix} + y^2\begin{bmatrix} \sinh\sigma_A^2 & 0 \\ 0 & \cosh\sigma_A^2 \end{bmatrix} + e^{-\sigma_A^2}\begin{bmatrix} 0 & xy \\ xy & 0 \end{bmatrix}.$$

And use $\cos a = x/r$ and $\sin a = y/r$ to obtain the components of $\mathcal{E}\mathbf{X}'\mathbf{X}'^T$ as

$$\begin{aligned} \mathcal{E}X'X' &= (\sigma_R^2+r^2)e^{-\sigma_A^2}(\cos^2 a\cosh\sigma_A^2 + \sin^2 a\sinh\sigma_A^2) \\ \mathcal{E}Y'Y' &= (\sigma_R^2+r^2)e^{-\sigma_A^2}(\cos^2 a\sinh\sigma_A^2 + \sin^2 a\cosh\sigma_A^2) \\ \mathcal{E}X'Y' &= \mathcal{E}Y'X' = (\sigma_R^2+r^2)e^{-2\sigma_A^2}\cos a\sin a\,. \end{aligned} \tag{147}$$

Also, rewrite (136) as

$$\boldsymbol{\mu}_{X'}\boldsymbol{\mu}_{X'}^T = r^2e^{-\sigma_A^2}\begin{bmatrix} \cos^2 a & \cos a\sin a \\ \cos a\sin a & \sin^2 a \end{bmatrix}. \tag{148}$$

Substitution of these last two expressions into $\mathbf{\Sigma}_{X'} = \mathcal{E}\mathbf{X}'\mathbf{X}'^T - \boldsymbol{\mu}_{X'}\boldsymbol{\mu}_{X'}{}^T$ yields

$$\begin{aligned} (\Sigma_{X'})_{xx}(r,a) &= \sigma_R^2e^{-\sigma_A^2}(\cos^2 a\cosh\sigma_A^2 + \sin^2 a\sinh\sigma_A^2) \\ &\quad + r^2e^{-\sigma_A^2}[\cos^2 a(\cosh\sigma_A^2 - 1) + \sin^2 a\sinh\sigma_A^2] \\ (\Sigma_{X'})_{xy}(r,a) &= (\Sigma_{X'})_{yx}(r,a) = [\sigma_R^2 + r^2(1-e^{+\sigma_A^2})]e^{-2\sigma_A^2}\cos a\sin a \\ (\Sigma_{X'})_{yy}(r,a) &= \sigma_R^2e^{-\sigma_A^2}(\cos^2 a\sinh\sigma_A^2 + \sin^2 a\cosh\sigma_A^2) \\ &\quad + r^2e^{-\sigma_A^2}[\cos^2 a\sinh\sigma_A^2 + \sin^2 a(\cosh\sigma_A^2 - 1)]\,. \end{aligned} \tag{149}$$

Note that (149) was given in [2]. Here its derivation has been provided, via (144).

4.2 The Estimation Errors of the PLKF and CLKF

The sample mean basically provides an unbiased estimate of the expected value of X', and the sample covariance matrix gives an estimate of the covariance matrix of X' [3]. But X' is biased as an estimator of $\mathbf{x}$. Indeed, by definition, the bias vector and mean-squared error matrix of X' as an estimator of $\mathbf{x}$ are respectively

$$\mathbf{b}_{X'} \equiv \mathcal{E}(X' - \mathbf{x}) \text{ and } \operatorname{mse} X' \equiv \mathcal{E}(X' - \mathbf{x})(X' - \mathbf{x})^T . \tag{150}$$

Obviously, since $\mathcal{E}X' = \boldsymbol{\mu}_{X'}$,

$$\mathbf{b}_{X'} = \boldsymbol{\mu}_{X'} - \mathbf{x} \tag{151}$$

And, since $X' - \mathbf{x} = X' - \boldsymbol{\mu}_{X'} + \boldsymbol{\mu}_{X'} - \mathbf{x} = X' - \boldsymbol{\mu}_{X'} + \mathbf{b}_{X'}$,

$$\operatorname{mse} X' = \boldsymbol{\Sigma}_{X'} + \mathbf{b}_{X'}\mathbf{b}_{X'}^T . \tag{152}$$

Also, using $\boldsymbol{\Sigma}_{X'} = \mathcal{E}X'X'^T - \boldsymbol{\mu}_{X'}\boldsymbol{\mu}_{X'}{}^T$,

$$\operatorname{mse} X' = \mathcal{E}X'X'^T + \mathbf{b}_{X'}\mathbf{b}_{X'}^T - \boldsymbol{\mu}_{X'}\boldsymbol{\mu}_{X'}{}^T . \tag{153}$$

Thus, when (133) is valid the bias vector of X' as an estimator of $\mathbf{x}$ is

$$\mathbf{b}_{X'} = (e^{-\sigma_A^2/2} - 1)\mathbf{x} . \tag{154}$$

And, using (153) with

$$\mathbf{b}_{X'}\mathbf{b}_{X'}^T - \boldsymbol{\mu}_{X'}\boldsymbol{\mu}_{X'}{}^T = (1 - e^{-\sigma_A^2/2})^2 \mathbf{x}\mathbf{x}^T - e^{-\sigma_A^2}\mathbf{x}\mathbf{x}^T = (1 - 2e^{-\sigma_A^2/2})\mathbf{x}\mathbf{x}^T ,$$

the mean squared error matrix of X' as an estimator of $\mathbf{x}$ is determined to be

$$\operatorname{mse} X' = (\sigma_R^2 + r^2)[\mathbf{I}(1 - e^{-2\sigma_A^2})/2 + \mathbf{x}\mathbf{x}^T (e^{-2\sigma_A^2})/r^2] + \mathbf{x}\mathbf{x}^T(1 - 2e^{-\sigma_A^2/2}) . \tag{155}$$

4.2.1 The Errors of the PLKF

Given the above bias vector and mean-squared error matrix of X' as an estimator of $\mathbf{x}$, the corresponding ones of the PLKF are determined as follows. (For now, just the equi-weighted scalar-weight case shall be used, and the superscripts "(s)" will be dropped.)

First, consider the sequence of estimates, $\hat{\mathbf{x}}'_n = (1/n)\sum_{m=1}^{n} \bar{\mathbf{x}}'_m$; and let $\bar{X}'_n$ and $\hat{X}'(n)$, $n = 1, 2, \cdots, N$, respectively denote the underlying random vectors of $\bar{\mathbf{x}}'_n$ and $\hat{\mathbf{x}}'_n$. The mean and variance of the $\hat{X}'(n)$ are [3]

$$\boldsymbol{\mu}_{\hat{X}'(n)} = \boldsymbol{\mu}_{X'} \text{ and } \boldsymbol{\Sigma}_{\hat{X}'(n)} = \boldsymbol{\Sigma}_{X'}/n \tag{156}$$

(since the underlying random vectors of the radar measurements are mutually independent, the $\bar{X}'_n = \mathbf{h}^{-1}(\bar{\boldsymbol{R}}_n)$'s are also mutually independent). Thus, using (152) as an identity, the bias vector and mean-squared error matrix of $\hat{X}'(n)$ as an estimator of $\mathbf{x}$ are

$$\mathbf{b}_{\hat{X}'(n)} = \mathbf{b}_{X'} \text{ and } \operatorname{mse}\hat{X}'(n) = \boldsymbol{\Sigma}_{X'}/n + \mathbf{b}_{X'}\mathbf{b}_{X'}^T. \tag{157}$$

Note that the PLKF remains biased as an estimator of $\mathbf{x}$ when $n \to \infty$.

4.2.2 The Errors of the C-LKF

Next, consider the C-LKF as an estimator of $\mathbf{x}$. And let $\hat{\boldsymbol{R}}(n)$ denote the underlying random vector of $\hat{\mathbf{r}}_n = (1/n)\sum_{m=1}^{n} \bar{\mathbf{r}}_m$. The corresponding random vector of the C-LKF is then $\hat{X}(n) = \mathbf{h}^{-1}(\hat{\boldsymbol{R}}(n))$. But since $\hat{\boldsymbol{R}}(n)$ is distributed as $\mathcal{N}(\mathbf{r}; \boldsymbol{\Sigma}_R/n)$ [3], the distribution of $\hat{X}(n)$ has the same form as the distribution of X', except with σ_R^2 and σ_A^2 replaced by $\sigma_{R(n)}^2 = \sigma_R^2/n$ and $\sigma_{A(n)}^2 = \sigma_A^2/n$. In particular, given (133) and (144),

$$\boldsymbol{\mu}_{\hat{X}(n)} = e^{-\sigma_A^2/2n}\mathbf{x} \tag{158}$$

and

$$\boldsymbol{\Sigma}_{\hat{X}(n)} = \left(r^2 + \sigma_R^2/n\right)\left\{\frac{1 - e^{-2\sigma_A^2/n}}{2}\begin{bmatrix} 1 & 0 \\ 0 & 1 \end{bmatrix} + \frac{e^{-2\sigma_A^2/n}}{r^2}\begin{bmatrix} x^2 & xy \\ xy & y^2 \end{bmatrix}\right\} - e^{-\sigma_A^2/n}\begin{bmatrix} x^2 & xy \\ yx & y^2 \end{bmatrix}. \tag{159}$$

Therefore, the estimation bias vector and mean-squared error matrix of the C-LKF as an estimator of $\mathbf{x}$ are

$$\mathbf{b}_{\hat{X}(n)} = (e^{-\sigma_A^2/2n} - 1)\mathbf{x}$$

and

$$\operatorname{mse} X' = (r^2 + \sigma_R^2/n)[\mathbf{I}(1 - e^{-2\sigma_A^2/n})/2 + \mathbf{x}\mathbf{x}^T(e^{-2\sigma_A^2/n})/r^2] + \mathbf{x}\mathbf{x}^T(1 - 2e^{-\sigma_A^2/2n}). \quad (160)$$

Note that, like the PLKF, the C-LKF is a biased estimator of $\mathbf{x}$. But, unlike the PLKF, the C-LKF is unbiased as $n \to \infty$.

4.3 The Popular PLKF "Debiasing" Methods

In the seminal paper on the Debiased Consistent Converted Measurements (DCCM) method [20] it was proposed that the rectangular pseudo-measurements be first "debiased" before being used in a PLKF. And there an *additive* correction was specified. Later (for reasons to be given below) a *multiplicative* one was recommended by others [4, 5], called the Unbiased Consistent Converted Measurements (UCCM) method. Those DCCM and UCCM proponents also recommended that the true covariance matrix of the rectangular pseudo-measurement be used, $\mathbf{\Sigma}_{X'}$ instead of $\mathbf{J}^{-1}(\overline{\mathbf{r}})\mathbf{\Sigma}_R\mathbf{J}^{-T}(\overline{\mathbf{r}})$.

The DCCM and the UCCM "debiasing" operations are basically

$$\overline{\mathbf{x}}_n'(\mathbf{b}) \equiv \overline{\mathbf{x}}_n' - \mathbf{b}_{X'} \quad \text{and} \quad \overline{\mathbf{x}}_n'(\lambda) \equiv (1/\lambda)\overline{\mathbf{x}}_n' \quad (161)$$

(here the argument "$\mathbf{b}$" serves to denote the DCCM case, and the argument "λ" serves to denote the UCCM case). And, using (161), the "debiased" PLKF estimates are

$$\hat{\mathbf{x}}_n^{\prime(s)}(\mathbf{b}) \equiv \frac{1}{n}\sum_{m=1}^{n}\overline{\mathbf{x}}_m'(\mathbf{b}) \quad \text{and} \quad \hat{\mathbf{x}}_n^{\prime(s)}(\lambda) \equiv \frac{1}{n}\sum_{m=1}^{n}\overline{\mathbf{x}}_m'(\lambda). \quad (162)$$

Figure 4-1 illustrates the effectiveness of the DCCM and UCCM methods for the scalar-weight case, using $\sigma_A = \pi/6$ in $\mathbf{b}_{X'} = (\lambda - 1)\mathbf{x}$ and $\lambda = e^{-\sigma_A^2/2}$. (Here the same measurement sets of the previous Chapter are being reused, partitioned into 50 mutually independent Monte Carlo trials.) And Figure 4-2 provides the sample means and "confidence intervals." Note that the corresponding sample means are mostly indistinguishable; and that the UCCM standard deviations are slightly larger than the DCCM ones – given (161), the true covariance matrices of the "debiased" measurements in the DCCM and UCCM cases are respectively $\mathbf{\Sigma}_{X'}$ and $(1/\lambda^2)\mathbf{\Sigma}_{X'}$, where $\lambda = e^{-\sigma_A^2/2}$.

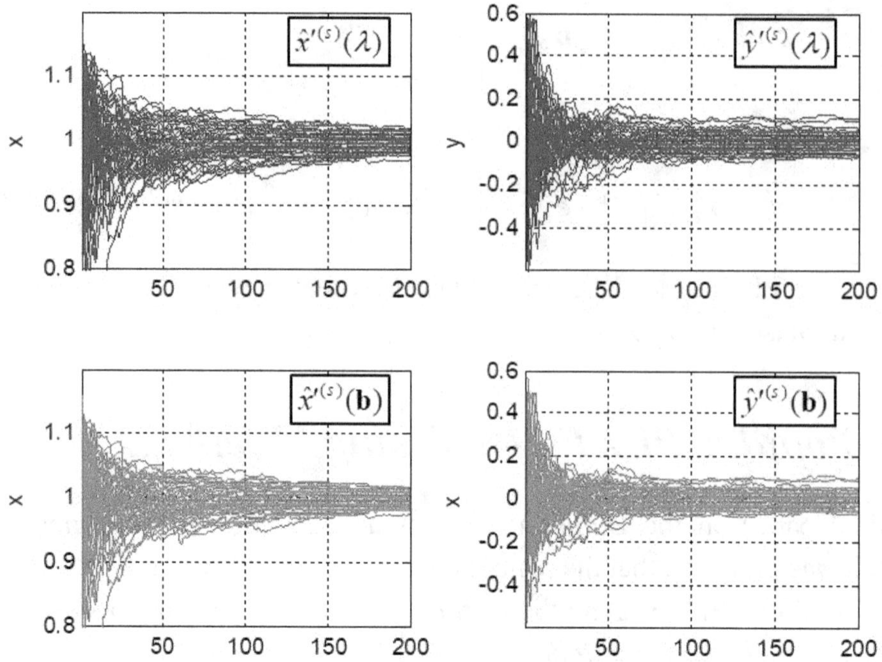

Figure 4-1 "Debiased" scalar-weight PLKF estimates

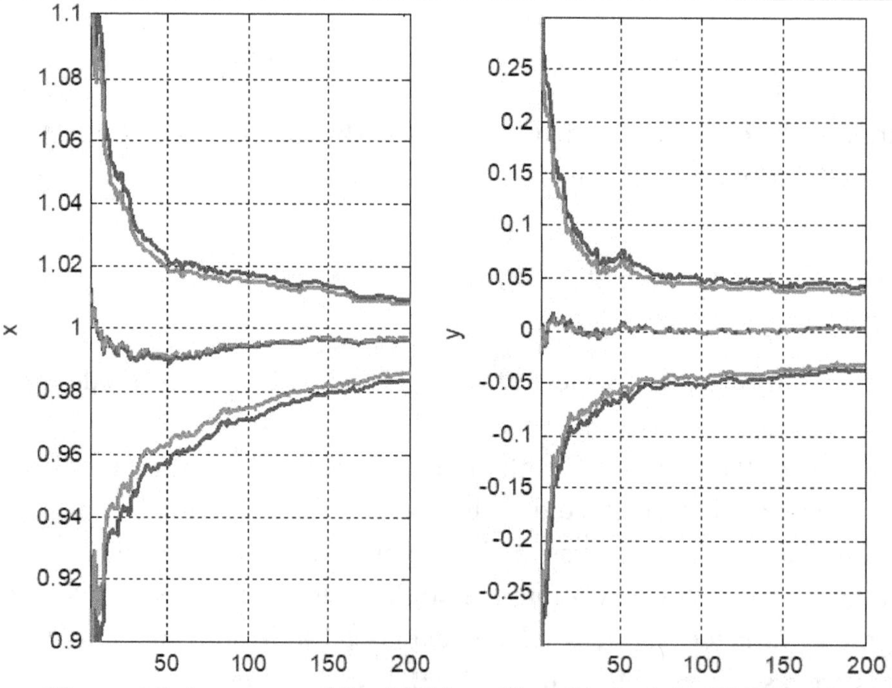

Figure 4-2 Accuracy of the "debiased" scalar-weight PLKF estimates

Unfortunately, such simplistic "debiasing" can lead to worse results in the matrix-weight case. Recall the definition of the matrix-weight PLKF:

$$\hat{\mathbf{x}}_n'^{(m)} = \hat{\mathbf{X}}_n'^{(m)} \sum_{m=1}^{n} \overline{\mathbf{X}}_m'^{-1} \overline{\mathbf{x}}_m' \text{ and } \hat{\mathbf{X}}_n'^{(m)} = \left(\sum_{m=1}^{n} \overline{\mathbf{X}}_m'^{-1} \right)^{-1}, \tag{163}$$

where $\overline{\mathbf{x}}_n' = \mathbf{h}^{-1}(\overline{\mathbf{r}}_n)$ and $\overline{\mathbf{X}}_n' = \mathbf{J}^{-1}(\overline{\mathbf{r}}_n)\mathbf{\Sigma}_R \mathbf{J}^{-T}(\overline{\mathbf{r}}_n)$. If $\overline{\mathbf{x}}_n'(\mathbf{b}) = \overline{\mathbf{x}}_n' - \mathbf{b}_{X'}$ is used in $\overline{\mathbf{X}}_n'$, that is,

$$\overline{\mathbf{X}}_n'(\mathbf{b}) \equiv \mathbf{J}^{-1}\left(\overline{\mathbf{x}}_n'(\mathbf{b})\right)\mathbf{\Sigma}_R \mathbf{J}^{-T}\left(\overline{\mathbf{x}}_n'(\mathbf{b})\right), \tag{164}$$

then

$$\hat{\mathbf{x}}_n'^{(m)}(\mathbf{b}) = \hat{\mathbf{X}}_n'^{(m)}(\mathbf{b}) \sum_{m=1}^{n} \overline{\mathbf{X}}_m'^{-1}(\mathbf{b}) \overline{\mathbf{x}}_m'(\mathbf{b}) \tag{165}$$

and

$$\hat{\mathbf{X}}_n'^{(m)}(\mathbf{b}) = \left(\sum_{m=1}^{n} \overline{\mathbf{X}}_m'^{-1}(\mathbf{b}) \right)^{-1}. \tag{166}$$

And if $\overline{\mathbf{x}}_n'(\lambda) \equiv (1/\lambda)\overline{\mathbf{x}}_n'$ is used in $\overline{\mathbf{X}}_n'$, that is,

$$\overline{\mathbf{X}}_n'(\lambda) \equiv \mathbf{J}^{-1}\left(\overline{\mathbf{x}}_n'(\lambda)\right)\mathbf{\Sigma}_R \mathbf{J}^{-T}\left(\overline{\mathbf{x}}_n'(\lambda)\right), \tag{167}$$

then

$$\hat{\mathbf{x}}_n'^{(m)}(\lambda) = \hat{\mathbf{X}}_n'^{(m)}(\lambda) \sum_{m=1}^{n} \overline{\mathbf{X}}_m'^{-1}(\lambda) \overline{\mathbf{x}}_m'(\lambda) \tag{168}$$

and

$$\hat{\mathbf{X}}_n'^{(m)}(\lambda) = \left(\sum_{m=1}^{n} \overline{\mathbf{X}}_m'^{-1}(\lambda) \right)^{-1}. \tag{169}$$

Figure 4-3 illustrates the sample means for the two "debiased" matrix-weight PLKF cases, respectively defined by (165) and (166), and by (167) and (168). (The corresponding C-LKF estimates are also shown for reference).

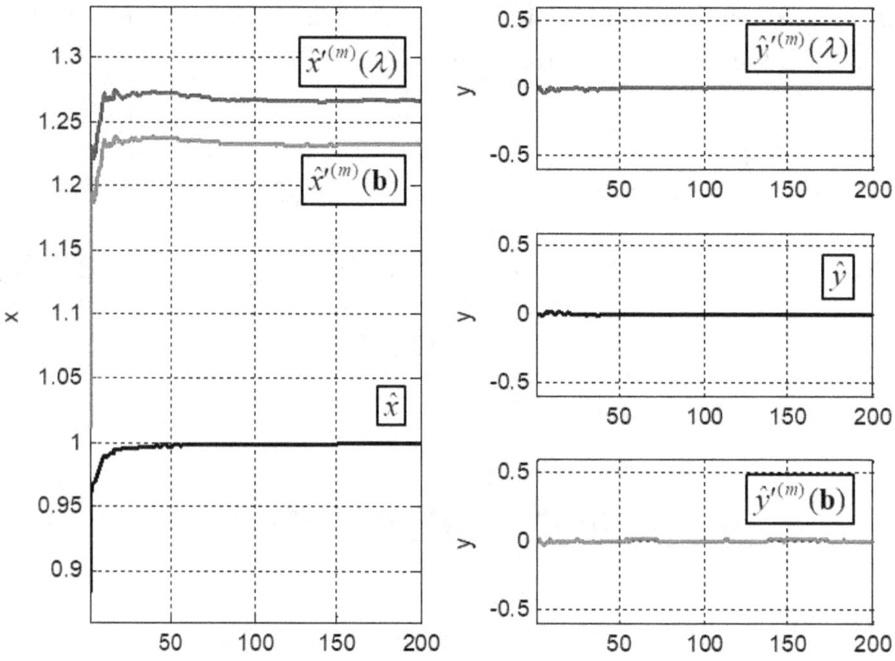

Figure 4-3 "Debiased" matrix-weight PLKF biases

Of course, the DCCM and UCCM proponents also say that the true covariance matrices should be used, $\mathbf{\Sigma}_{X'}$, not $\mathbf{J}^{-1}(\overline{\mathbf{r}}_n)\mathbf{\Sigma}_R\mathbf{J}^{-T}(\overline{\mathbf{r}}_n)$. Unfortunately, $\mathbf{\Sigma}_{X'}$ is a function of the unknown being estimated, $\mathbf{x}$, and so in practice some approximation must be used instead. For example: given $\overline{\mathbf{r}}$, use $\mathbf{\Sigma}_{X'}(\overline{\mathbf{r}})$; or, given $\hat{\mathbf{x}}$, use $\mathbf{\Sigma}_{X'}(\hat{\mathbf{x}})$. Here these two cases shall be distinguished by respectively calling them *measurement-based* and *estimate-based* PLKF's, written MB-PLKF and the EB-PLKF. (Note that since the basic PLKF uses $\mathbf{J}^{-1}(\overline{\mathbf{r}})\mathbf{\Sigma}_R\mathbf{J}^{-T}(\overline{\mathbf{r}})$, it too is an MB-PLKF.) More unfortunate, however, the $\mathbf{b}_{X'}$ that the DCCM uses also depends upon $\mathbf{x}$ – the *ideal* DCCM was actually being illustrated above, with $\mathbf{b}_{X'}=(\lambda-1)\mathbf{x}$. In contrast, the λ in the UCCM depends only on σ_A. And so only the more practical UCCM shall be considered further in the sequel.

Accordingly, let the MB-PLKF use $\overline{\mathbf{x}}'_n(\lambda)$ in $\mathbf{\Sigma}_{X'}$, $n=1,2,\cdots,N$, and determine

$$\hat{\mathbf{x}}'^{(m)}_n(\overline{\lambda})=\hat{\mathbf{X}}'^{(m)}_n(\overline{\lambda})\sum_{m=1}^{n}\mathbf{\Sigma}^{-1}_{X'}\left(\overline{\mathbf{x}}'_n(\lambda)\right)\overline{\mathbf{x}}'_m(\lambda) \tag{170}$$

and

$$\hat{\mathbf{X}}'^{(m)}_n(\bar{\lambda}) = \left(\sum_{m=1}^{n} \boldsymbol{\Sigma}^{-1}_{X'}(\bar{\mathbf{x}}'_n(\lambda))\right)^{-1}. \tag{171}$$

And, alternatively, let the EB-PLKF use $\hat{\mathbf{x}}'^{(m)}_{n-1}(\hat{\lambda})$ in $\boldsymbol{\Sigma}_{X'}$, $n = 2, 3, \cdots, N$, and determine

$$\hat{\mathbf{x}}'^{(m)}_n(\hat{\lambda}) = \hat{\mathbf{X}}'^{(m)}_n(\hat{\lambda})\left[\left(\hat{\mathbf{X}}'^{(m)}_{n-1}(\hat{\lambda})\right)^{-1}\hat{\mathbf{x}}'^{(m)}_{n-1}(\hat{\lambda}) + \boldsymbol{\Sigma}^{-1}_{X'}(\hat{\mathbf{x}}'^{(m)}_{n-1}(\hat{\lambda}))\bar{\mathbf{x}}'_m(\lambda)\right] \tag{172}$$

and

$$\hat{\mathbf{X}}'^{(m)}_n(\hat{\lambda}) = \left[\left(\hat{\mathbf{X}}'^{(m)}_{n-1}(\hat{\lambda})\right)^{-1} + \boldsymbol{\Sigma}^{-1}_{X'}(\hat{\mathbf{x}}'^{(m)}_{n-1}(\hat{\lambda}))\right]^{-1}, \tag{173}$$

with $\hat{\mathbf{x}}'^{(m)}_1(\hat{\lambda}) = \hat{\mathbf{x}}'^{(m)}_1(\bar{\lambda})$ and $\hat{\mathbf{X}}'^{(m)}_1(\hat{\lambda}) = \hat{\mathbf{X}}'^{(m)}_1(\bar{\lambda})$. [Between (170) and (171), the common factor $(1/\lambda^2)$ has been tacitly canceled – also between (172) and (173).] Figure 4-4 shows these two "debiased and consistent" PLKF cases, along with "biased" C-LKF case.

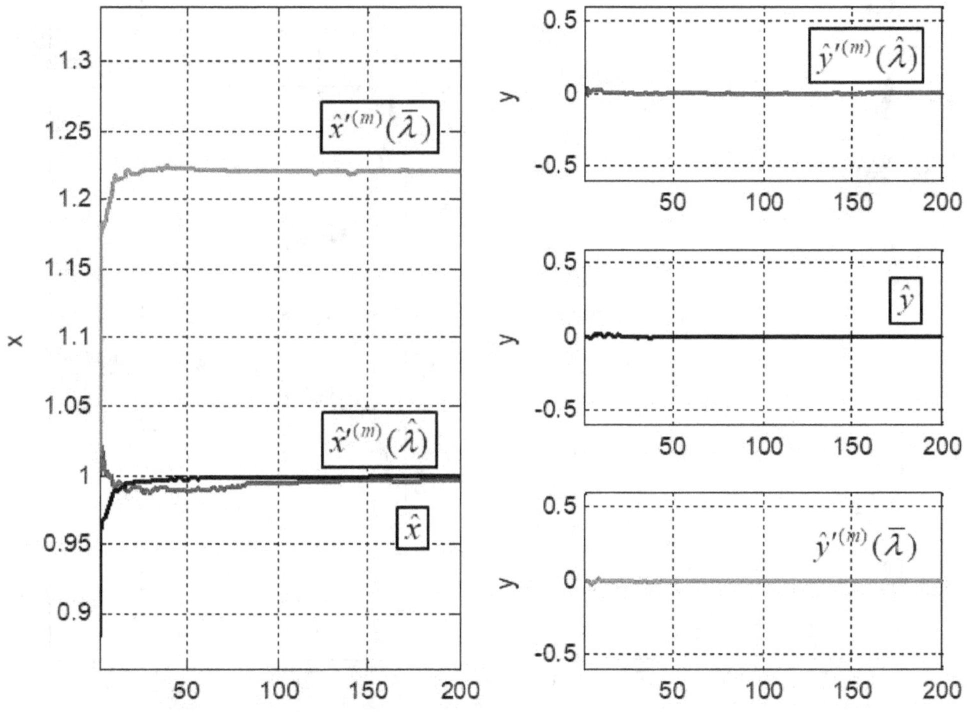

Figure 4-4 Matrix-weight UCCM with "consistent" covariance matrices

4.4 Discussion of the "debiased" PLKF

The above presentation is summarized as follows. The PLKF is a biased estimator of $\mathbf{x}$ as $N \to \infty$, while the C-LKF is asymptotically unbiased. The (ideal) DCCM "debiases" the PLKF by subtracting the true estimation bias from the (unbiased) rectangular pseudo-measurements; and the (practical) UCCM "debiases" the PLKF by making those measurements noisier. And both "debiased" PLKF's are much noisier than the C-LKF.

Figure 4-5 shows the sample means and "confidence intervals" for the estimates that were shown earlier in Figure 4-2, along with the true standard deviations of the DCCM and UCCM cases – the dashed curves. There the DCCM standard deviations are seen to be smaller than those of the UCCM – the DCCM covariance matrix is $\mathbf{\Sigma}_{X'}$, given by (144) , while the UCCM covariance matrix is $(1/\lambda^2)\mathbf{\Sigma}_{X'}$. The corresponding curves for the C-LKF are also shown in the figure (the darker curves).

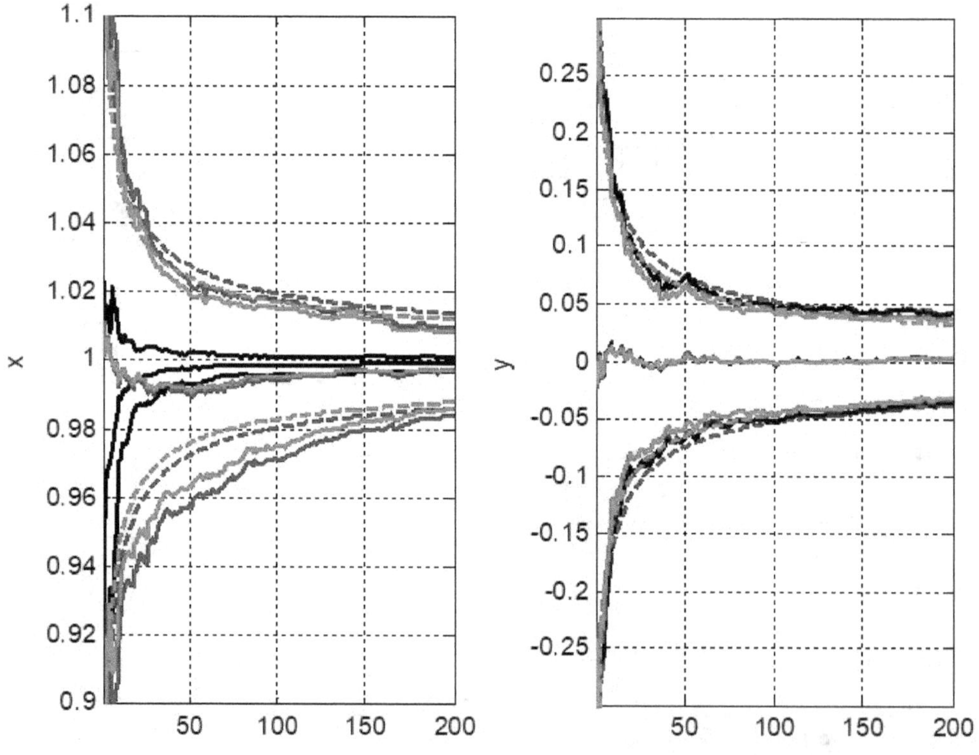

Figure 4-5 Accuracies of the "debiased" PLKF's (scalar-weight cases)

4.5 *The Biases of the "Debiased and Consistent" PLKF*

Unfortunately, the "debiased and consistent" matrix-weight PLKF's have residual biases. The reason is the same as why the bias of the matrix-weight PLKF has the opposite sense to the bias of the scalar-weight PLKF.

Consider the basic matrix-weight PLKF, but written as

$$\hat{\mathbf{X}}_n'^{-1}\hat{\mathbf{x}}_n' = \sum_{m=1}^{n} \overline{\mathbf{X}}_m'^{-1}\overline{\mathbf{x}}_m' \quad \text{and} \quad \hat{\mathbf{X}}_n'^{-1} = \sum_{m=1}^{n} \overline{\mathbf{X}}_m'^{-1} . \tag{174}$$

Dropping the adornments, the functional form of the product $\mathbf{X}^{-1}\mathbf{x} = \mathbf{J}^T(\mathbf{x})\mathbf{\Sigma}_R^{-1}\mathbf{J}(\mathbf{x})\mathbf{h}^{-1}(\mathbf{r})$ is determined as follows. First, the component form of the product $\mathbf{J}(\mathbf{x})\mathbf{h}^{-1}(\mathbf{r})$ is

$$\begin{bmatrix} \cos a & \sin a \\ -\sin a/r & \cos a/r \end{bmatrix} \begin{bmatrix} r\cos a \\ r\sin a \end{bmatrix} = \begin{bmatrix} r \\ 0 \end{bmatrix}. \tag{175}$$

And, given $\mathbf{\Sigma}_R$, symmetric and positive definite, the component form of $\mathbf{\Sigma}_R^{-1}\mathbf{J}\mathbf{h}^{-1}$ is

$$\begin{bmatrix} 1/\sigma_R^2 & 0 \\ 0 & 1/\sigma_A^2 \end{bmatrix} \begin{bmatrix} \cos a & \sin a \\ -\sin a/r & \cos a/r \end{bmatrix} \begin{bmatrix} r\cos a \\ r\sin a \end{bmatrix} = \begin{bmatrix} r/\sigma_R^2 \\ 0 \end{bmatrix}. \tag{176}$$

Multiplying this result by $\mathbf{J}^T$ yields

$$\begin{bmatrix} \cos a & -\sin a/r \\ \sin a & \cos a/r \end{bmatrix} \begin{bmatrix} r/\sigma_R^2 \\ 0 \end{bmatrix} = \frac{1}{\sigma_R^2} \begin{bmatrix} r\cos a \\ r\sin a \end{bmatrix}. \tag{177}$$

Thus,

$$\mathbf{X}^{-1}\mathbf{x} = (1/\sigma_R^2)\mathbf{x} . \tag{178}$$

Note that in (178) the *effective* weight of $\mathbf{X}^{-1}\mathbf{x}$ is independent of σ_A^2. But $\mathbf{X}^{-1}$ is

$$(\mathbf{J}^{-1}\mathbf{\Sigma}_R\mathbf{J}^{-T})^{-1} = \mathbf{J}^T\mathbf{\Sigma}_R^{-1}\mathbf{J} = \mathbf{O}^T\mathbf{D}^{-1}\mathbf{\Sigma}_R^{-1}\mathbf{D}^{-1}\mathbf{O} \tag{179}$$

– see (17) and (18) in Chapter 2. And its component form is

$$\mathbf{X}^{-1} = \begin{bmatrix} \cos^2 a/\sigma_R^2 + \sin^2 a/(r\sigma_A)^2 & +\cos a \sin a/\sigma_R^2 - \cos a \sin a/(r\sigma_A)^2 \\ +\cos a \sin a/\sigma_R^2 - \cos a \sin a/(r\sigma_A)^2 & \sin^2 a/\sigma_R^2 + \cos^2 a/(r\sigma_A)^2 \end{bmatrix},$$

which, for convenience, is written as

$$\mathbf{X}^{-1} = \mathbf{X}_R^{-1} + \mathbf{X}_A^{-1},$$

with

$$\mathbf{X}_R^{-1} \equiv \frac{1}{\sigma_R^2}\begin{bmatrix} \cos^2 a & \cos a \sin a \\ \cos a \sin a & \sin^2 a \end{bmatrix}$$

and

$$\mathbf{X}_A^{-1} \equiv \frac{1}{r^2\sigma_A^2}\begin{bmatrix} \sin^2 a & -\cos a \sin a \\ -\cos a \sin a & \cos^2 a \end{bmatrix}$$

(the notation here is a contrivance – both $\mathbf{X}_R^{-1}$ and $\mathbf{X}_A^{-1}$ are singular). Thus the effective weight matrix of $\mathbf{X}^{-1}\mathbf{x}$ is inconsistent with the weight of $\mathbf{X}^{-1}$.

Consider the recursive update form of the basic MB-PLKF (without "debiasing"),

$$\hat{\mathbf{x}}'_n = \hat{\mathbf{x}}'_{n-1} + \hat{\mathbf{X}}'_n \bar{\mathbf{X}}_n'^{-1}\left(\bar{\mathbf{x}}'_n - \hat{\mathbf{x}}'_{n-1}\right), \tag{180}$$

where

$$\bar{\mathbf{X}}'_n \equiv \bar{\mathbf{X}}'(\bar{\mathbf{r}}_n) = \mathbf{J}^{-1}(\bar{\mathbf{r}}_n)\boldsymbol{\Sigma}_R \mathbf{J}^{-T}(\bar{\mathbf{r}}_n). \tag{181}$$

Given (178) with $\bar{\mathbf{X}}'_n = \bar{\mathbf{X}}'(\bar{\mathbf{r}}_n)$, the weighted measurement is $\bar{\mathbf{X}}_n'^{-1}\bar{\mathbf{x}}'_n = (1/\sigma_R^2)\bar{\mathbf{x}}'_n$, and so in (180) the product of the gain matrix and the measurement, $\hat{\mathbf{X}}'_n\bar{\mathbf{X}}_n'^{-1}\bar{\mathbf{x}}'_n$, is $\hat{\mathbf{X}}'_n\bar{\mathbf{x}}'_n/\sigma_R^2$. But the product of the MB-gain matrix and the prior estimate is $\hat{\mathbf{X}}'_n\bar{\mathbf{X}}_n'^{-1}\hat{\mathbf{x}}'_{n-1}$. Similarly, in the basic EB-PLKF, since $\bar{\mathbf{X}}'_n \equiv \bar{\mathbf{X}}'(\hat{\mathbf{x}}'_{n-1}) = \mathbf{J}^{-1}(\hat{\mathbf{x}}'_{n-1})\boldsymbol{\Sigma}_R\mathbf{J}^{-T}(\hat{\mathbf{x}}'_{n-1})$, the product of the gain matrix and the prior estimate is $\hat{\mathbf{X}}'_n\hat{\mathbf{x}}'_{n-1}/\sigma_R^2$, while the product of the gain matrix and the measurement is $\hat{\mathbf{X}}'_n\bar{\mathbf{X}}_n'^{-1}\bar{\mathbf{x}}'_{n-1}$. Thus, in both the MB-PLKF and EB-PLKF, the effective gains that respectively operate on the given measurement and the prior estimate are "inconsistent" because of the functional dependency between the realization and its transformed associated covariance matrix.

Fortunately, a simple remedy for the unmodeled correlation problem exists when $n>1$: instead of $\bar{\mathbf{X}}'(\bar{\mathbf{x}}'_n) = \mathbf{J}^{-1}(\bar{\mathbf{x}}'_n)\boldsymbol{\Sigma}_R\mathbf{J}^{-T}(\bar{\mathbf{x}}'_n)$, simply use

$$\bar{\mathbf{X}}'_n \equiv \bar{\mathbf{X}}'(\bar{\mathbf{x}}'_{n-1}) = \mathbf{J}^{-1}(\bar{\mathbf{x}}'_{n-1})\boldsymbol{\Sigma}_R\mathbf{J}^{-T}(\bar{\mathbf{x}}'_{n-1}). \tag{182}$$

That is, let

$$\hat{\mathbf{x}}_n' = \hat{\mathbf{X}}_n' \left\{ \left[\overline{\mathbf{X}}'(\overline{\mathbf{x}}_1') \right]^{-1} \overline{\mathbf{x}}_1' + \sum_{m=2}^{n} \left[\overline{\mathbf{X}}'(\overline{\mathbf{x}}_{m-1}') \right]^{-1} \overline{\mathbf{x}}_m' \right\} \tag{183}$$

and

$$\hat{\mathbf{X}}_n'^{-1} = \left[\overline{\mathbf{X}}'(\overline{\mathbf{x}}_1') \right]^{-1} + \sum_{m=2}^{n} \left[\overline{\mathbf{X}}'(\overline{\mathbf{x}}_{m-1}') \right]^{-1}. \tag{184}$$

In which case the effective gains used on the measurement and prior estimate become "consistent" as n increases – for small n there is still a problem, because $\hat{\mathbf{x}}_1 = \overline{\mathbf{x}}_1'$.

Figure 4-6 illustrates the effectiveness of using (182) in the basic PLKF matrix-weight update, along with the basic scalar-weight case. Also shown are $\mu_{X'}$ and $\mu_{Y'}$, the expected values of the rectangular pseudo-measurements (the dark lines). Figure 4-7 provides the corresponding sample means and "confidence intervals" (there the solid curves are the matrix-weight case, and the dashed curves are the scalar-weight case). This shows that the remedy given by (182) comes at a cost: the estimates appear to be noisier. Obviously, since in (183), when $m > 1$ each summand is the product of a random matrix and an independent random vector. Recall that the variance of a product of mutually independent random variables is the sum of the variances of the respective variables. And so these PLKF estimates are noisier than the basic matrix-weight ones.

The functional dependency between certain realizations and their transformed associated covariance matrices in the MB-PLKF and EB-PLKF cases is a correlation that is not modeled in the DCCM and UCCM methods. In the "debiased and consistent" MB-PLKF UCCM case $\boldsymbol{\Sigma}_{X'}^{-1}(\overline{\mathbf{x}}_n'(\lambda))$ and $\overline{\mathbf{x}}_m'(\lambda)$ are correlated; and in the "debiased and consistent" EB-PLKF UCCM case $\boldsymbol{\Sigma}_{X'}^{-1}(\hat{\mathbf{x}}_{n-1}'^{(m)}(\hat{\lambda}))$ and $\overline{\mathbf{x}}_m'(\lambda)$ are correlated. But, using the above remedy, $\overline{\mathbf{X}}'^{-1}(\overline{\mathbf{x}}_{n-1}') = \boldsymbol{\Sigma}_{X'}^{-1}(\overline{\mathbf{x}}_{n-1}'(\lambda))$, the effective gains that operate upon the measurement and prior estimate can become the same (as n increases). Of course, this (asymptotic) remedy comes at a cost: the UCCM estimates, which were already noisier by $1/\lambda$, are made even noisier. In the sequel the *prior-weight* "debiased" and now consistent UCCM measurement-based case shall be used for the illustrations.

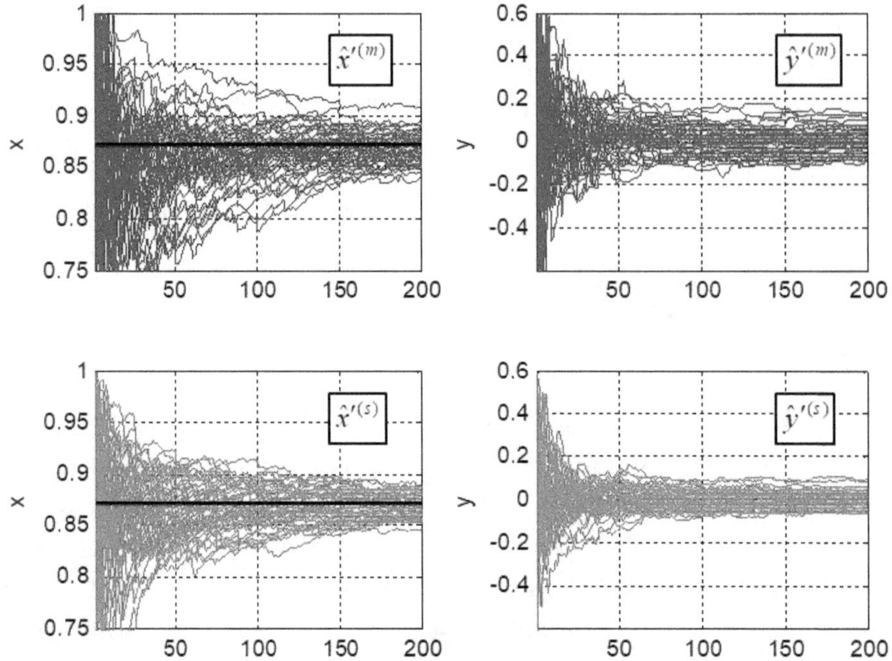

Figure 4-6 Comparison of the "prior-weight" MB-PLKF with the basic scalar-weight PLKF

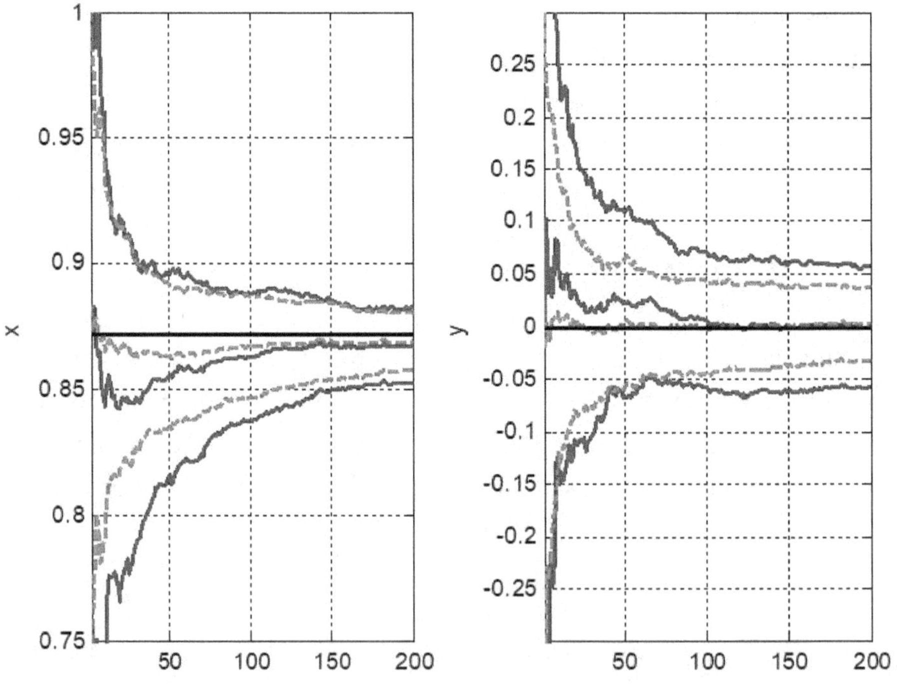

Figure 4-7 Accuracies of prior-matrix weight and scalar-weight PLKF's

4.6 Chapter 4 References

[1] K. S. Miller, Engineering Mathematics, Dover (1963).

[2] D. Lerro and T. Bar-Shalom, "Tracking with Debiased Consistent Converted Measurements versus EKF," in Transactions on Aerospace and Electronic Systems, IEEE AES-29, pp. 1015-1022 (1993).

[3] B. V. Gnedenko, Probability Theory, CRC Press (1998).

[4] L. Mo, X. Song, Y. Zhou, and Z. Sun, "Alternative Unbiased Consistent Converted Measurements for Target Tracking," in Proceedings of Acquisition, Tracking, and Pointing XI, M. K. Masten and L. A. Stockum (eds.), SPIE Vol. 3086, p. 308-310 (1997).

[5] L. Mo, X. Song, Y. Zhou, Z. Sun, and Y. Bar-Shalom, "Unbiased Converted Measurements for Tracking," in Transactions on Aerospace and Electronic Systems, IEEE AES-34, pp. 1023-1026 (1998).

5

Analysis of the EKF Update

In Chapter 3 it was shown that the EKF can perform better than a PLKF. Here the EKF is discussed more formally; and some analytic results are obtained for use in the sequel.

In this Chapter the alternate form of the EKF update equations shall be used, and for convenience the primes and indices on $\hat{\mathbf{x}}_n''$ and $\hat{\mathbf{X}}_n''$ will be dropped. That is, given a prior estimate of $\mathbf{x}$, and its associated covariance matrix, written $(\hat{\mathbf{x}}^-;\hat{\mathbf{X}}^-)$, the EKF update equations will be written as

$$\hat{\mathbf{x}} = \hat{\mathbf{x}}^- + \mathbf{K}\left[\bar{\mathbf{r}} - \mathbf{h}(\hat{\mathbf{x}}^-)\right] \text{ and } \hat{\mathbf{X}} = \left[(\hat{\mathbf{X}}^-)^{-1} + \mathbf{J}^T(\hat{\mathbf{x}}^-)\boldsymbol{\Sigma}_R^{-1}\mathbf{J}(\hat{\mathbf{x}}^-)\right]^{-1}, \tag{185}$$

with

$$\mathbf{K} = \hat{\mathbf{X}}\mathbf{J}^T(\hat{\mathbf{x}}^-)\boldsymbol{\Sigma}_R^{-1}. \tag{186}$$

As before, $\boldsymbol{R}$ is the underlying random vector of $\bar{\mathbf{r}}$, distributed as $\mathcal{N}(\mathbf{r};\boldsymbol{\Sigma}_R)$; and the underlying random variables of the range and azimuth measurements, R and A, are mutually independent. But now an underlying random vector for $\hat{\mathbf{x}}^-$ is also invoked, denoted $\boldsymbol{X}^-$. Its distribution is arbitrary, except for its covariance matrix, which is assumed to exist and be positive definite; and $\boldsymbol{X}^-$ and $\boldsymbol{R}$ are tacitly assumed to be mutually independent.

Note that R and A mutually independent implies $\mathbf{\Sigma}_{\boldsymbol{R}} = \mathbf{\Sigma}_R + \mathbf{\Sigma}_A$; specifically,

$$\mathbf{\Sigma}_R \equiv \begin{bmatrix} \sigma_R^2 & 0 \\ 0 & 0 \end{bmatrix} \text{ and } \mathbf{\Sigma}_A \equiv \begin{bmatrix} 0 & 0 \\ 0 & \sigma_A^2 \end{bmatrix}, \tag{187}$$

and that

$$\overline{\mathbf{X}}_{\boldsymbol{R}} = \mathbf{J}^{-1}(\hat{\mathbf{r}}^-)\left(\mathbf{\Sigma}_R + \mathbf{\Sigma}_A\right)\mathbf{J}^{-T}(\hat{\mathbf{r}}^-) = \overline{\mathbf{X}}_R + \overline{\mathbf{X}}_A, \tag{188}$$

where

$$\overline{\mathbf{X}}_R \equiv \mathbf{J}^{-1}(\hat{\mathbf{r}}^-)\mathbf{\Sigma}_R\mathbf{J}^{-T}(\hat{\mathbf{r}}^-) \text{ and } \overline{\mathbf{X}}_A \equiv \mathbf{J}^{-1}(\hat{\mathbf{r}}^-)\mathbf{\Sigma}_A\mathbf{J}^{-T}(\hat{\mathbf{r}}^-). \tag{189}$$

5.1 The EKF Update Errors

Now given $(\hat{\mathbf{x}}^-;\hat{\mathbf{X}}^-)$ and $\boldsymbol{R}$, determine $\hat{\mathbf{X}}$ by the second expression in (185). And then define the conditional random vector

$$\boldsymbol{X}_{(\hat{\mathbf{x}}^-;\hat{\mathbf{X}}^-)} = \hat{\mathbf{x}}^- + \hat{\mathbf{X}}\mathbf{J}^T(\hat{\mathbf{x}}^-)\mathbf{\Sigma}_{\boldsymbol{R}}^{-1}\left[\boldsymbol{R} - \mathbf{h}(\hat{\mathbf{x}}^-)\right]. \tag{190}$$

Note that $\hat{\mathbf{X}}$ is conditionally deterministic: it depends only on $(\hat{\mathbf{x}}^-;\hat{\mathbf{X}}^-)$ and $\mathbf{\Sigma}_{\boldsymbol{R}}$. Thus, $\boldsymbol{X}_{(\hat{\mathbf{x}}^-;\hat{\mathbf{X}}^-)}$ is conditionally gaussian [1]. Indeed, $\boldsymbol{X}_{(\hat{\mathbf{x}}^-;\hat{\mathbf{X}}^-)} \sim \mathcal{N}(\boldsymbol{\mu}_{X|(\hat{\mathbf{x}}^-;\hat{\mathbf{X}}^-)};\mathbf{\Sigma}_{X|(\hat{\mathbf{x}}^-;\hat{\mathbf{X}}^-)})$, where

$$\boldsymbol{\mu}_{X|(\hat{\mathbf{x}}^-;\hat{\mathbf{X}}^-)} = \hat{\mathbf{x}}^- + \hat{\mathbf{X}}\mathbf{J}^T(\hat{\mathbf{x}}^-)\mathbf{\Sigma}_{\boldsymbol{R}}^{-1}\left[\boldsymbol{\mu}_{\boldsymbol{R}} - \mathbf{h}(\hat{\mathbf{x}}^-)\right] \tag{191}$$

and

$$\mathbf{\Sigma}_{X|(\hat{\mathbf{x}}^-;\hat{\mathbf{X}}^-)} = \hat{\mathbf{X}}\mathbf{J}^T(\hat{\mathbf{x}}^-)\mathbf{\Sigma}_{\boldsymbol{R}}^{-1}\mathbf{J}(\hat{\mathbf{x}}^-)\hat{\mathbf{X}}^T. \tag{192}$$

This last expression follows from $\boldsymbol{X}_{(\hat{\mathbf{x}}^-;\hat{\mathbf{X}}^-)} - \mathcal{E}\boldsymbol{X}_{(\hat{\mathbf{x}}^-;\hat{\mathbf{X}}^-)} = \hat{\mathbf{X}}\mathbf{J}^T(\hat{\mathbf{x}}^-)\mathbf{\Sigma}_{\boldsymbol{R}}^{-1}(\boldsymbol{R} - \mathcal{E}\boldsymbol{R})$.

Unfortunately, as an estimator of $\mathbf{x}$, (190) has an unmodeled linearization error. For example, consider the scalar-weight EKF update, defined here as

$$\hat{\mathbf{x}}^{(s)} = \hat{\mathbf{x}}^- + \frac{\overline{w}}{\hat{w}}\mathbf{J}^{-1}(\hat{\mathbf{r}}^-)\left(\overline{\mathbf{r}} - \mathbf{h}(\hat{\mathbf{x}}^-)\right) \text{ and } \hat{w} = \hat{w}^- + \overline{w}, \tag{193}$$

where $\hat{\mathbf{X}}^- \equiv \mathbf{I}/\hat{w}^-$ and $\overline{\mathbf{X}} \equiv \mathbf{I}/\overline{w}$. This update is a linear (affine) relation in $\overline{\mathbf{r}}$, with a nonlinear dependency upon $\hat{\mathbf{r}}^-$. Together with $\hat{\mathbf{r}}^- = \mathbf{h}(\hat{\mathbf{x}}^-)$, it leads to

$$\hat{w}\left(\hat{\mathbf{x}}^{(s)} - \hat{\mathbf{x}}^-\right) = \overline{w}\mathbf{J}^{-1}(\hat{\mathbf{r}}^-)\left(\overline{\mathbf{r}} - \hat{\mathbf{r}}^-\right), \tag{194}$$

a relation between weighted differentials.

Using $\hat{r}^- = [(\hat{x}^-)^2 + (\hat{y}^-)^2]^{1/2}$ and $\hat{a}^- = \arctan(\hat{x}^-, \hat{y}^-)$, in component form (193) is

$$\begin{bmatrix} \hat{x}^{(s)} \\ \hat{y}^{(s)} \end{bmatrix} = \begin{bmatrix} \hat{x}^- \\ \hat{y}^- \end{bmatrix} + \frac{\overline{w}}{\hat{w}} \begin{bmatrix} \cos\hat{a}^- & -\hat{r}^- \sin\hat{a}^- \\ +\sin\hat{a}^- & \hat{r}^- \cos\hat{a}^- \end{bmatrix} \begin{bmatrix} \overline{r} - \hat{r}^- \\ \overline{a} - \hat{a}^- \end{bmatrix}. \tag{195}$$

Equivalently,

$$\begin{bmatrix} \hat{x}^{(s)} \\ \hat{y}^{(s)} \end{bmatrix} = \begin{bmatrix} \hat{x}^- \\ \hat{y}^- \end{bmatrix} + \frac{r_\Delta}{\hat{r}^-} \begin{bmatrix} \hat{x}^- \\ \hat{y}^- \end{bmatrix} + a_\Delta \begin{bmatrix} -\hat{y}^- \\ +\hat{x}^- \end{bmatrix}, \tag{196}$$

where

$$r_\Delta \equiv \frac{\overline{w}}{\hat{w}}(\overline{r} - \hat{r}^-) \quad \text{and} \quad a_\Delta \equiv \frac{\overline{w}}{\hat{w}}(\overline{a} - \hat{a}^-). \tag{197}$$

Using $\hat{x}^{(s)} = \hat{r}^{(s)} \cos\hat{a}^{(s)}$ and $\hat{y}^{(s)} = \hat{r}^{(s)} \sin\hat{a}^{(s)}$ in (193),

$$\hat{r}^{(s)} \cos\hat{a}^{(s)} = (\hat{r}^- + r_\Delta)\cos\hat{a}^- - a_\Delta \hat{r}^- \sin\hat{a}^- \tag{198}$$

and

$$\hat{r}^{(s)} \sin\hat{a}^{(s)} = (\hat{r}^- + r_\Delta)\sin\hat{a}^- + a_\Delta \hat{r}^- \cos\hat{a}^-. \tag{199}$$

leads to

$$\hat{r}^{(s)} = \sqrt{(\hat{r}^- + r_\Delta)^2 + (a_\Delta \hat{r}^-)^2} \tag{200}$$

and

$$\hat{a}^{(s)} = \arctan\left[\frac{(\hat{r}^- + r_\Delta)\sin\hat{a}^- + a_\Delta \hat{r}^- \cos\hat{a}^-}{(\hat{r}^- + r_\Delta)\cos\hat{a}^- - a_\Delta \hat{r}^- \sin\hat{a}^-}\right]. \tag{201}$$

Now let $a_\Delta = 0$ in (200) and (201), with r_Δ arbitrary. In which case

$$\hat{r}^{(s)}_{\bar{a}=\hat{a}^-} = \hat{r}^- + r_\Delta \text{ and } \hat{a}^{(s)}_{\bar{a}=\hat{a}^-} = \hat{a}^- . \tag{202}$$

Note that this range update is what the (optimal) LKF would determine – see Chapter 3. More important, it does not change the azimuth estimate, which is appropriate since the mutual independence of R and A implies that $\bar{r}$ contains no information on a. Alternatively, let $r_\Delta = 0$ with a_Δ arbitrary. In which case (200) and (201) become

$$\hat{r}^{(s)}_{\bar{r}=\hat{r}^-} = \hat{r}^- \sqrt{1+a_\Delta^2} \text{ and } \hat{a}^{(s)}_{\bar{r}=\hat{r}^-} = \arctan\left(\frac{\hat{y}^- + a_\Delta \hat{x}^-}{\hat{x}^- - a_\Delta \hat{y}^-}\right). \tag{203}$$

Here the range component is affected, which is inappropriate since $\bar{a}$ contains no information on r. Indeed, in (203), the range update is seen to be approximately $\hat{r}^- + |a_\Delta \hat{r}^-|$; and, rewriting the second expression in (203) as

$$\hat{a}^{(s)}_{\bar{r}=\hat{r}^-} = \arctan\left(\frac{\sin\hat{a}^- + a_\Delta \cos\hat{a}^-}{\cos\hat{a}^- - a_\Delta \sin\hat{a}^-}\right), \tag{204}$$

and then using the identity $\alpha + \arctan b = \arctan[(\sin\alpha + b\cos\alpha)/(\cos\alpha - b\sin\alpha)]$, the azimuth update is seen to be approximately $\hat{a}^- + a_\Delta + a_\Delta^3/3$.

For the *matrix-weight* EKF case, use $\bar{\mathbf{X}}_R^{-1} \equiv \mathbf{J}^T(\hat{\mathbf{x}}^-)\boldsymbol{\Sigma}_R^{-1}\mathbf{J}(\hat{\mathbf{x}}^-)$ and the identity $\mathbf{J}^T(\hat{\mathbf{r}}^-)\boldsymbol{\Sigma}_R^{-1} \equiv \mathbf{J}^T(\hat{\mathbf{x}}^-)\boldsymbol{\Sigma}_R^{-1}\mathbf{J}(\hat{\mathbf{x}}^-)\mathbf{J}^{-1}(\hat{\mathbf{x}}^-)$ to rewrite the first expression in (185) as

$$\hat{\mathbf{x}}^{(m)} = \hat{\mathbf{x}}^- + \hat{\mathbf{X}}\bar{\mathbf{X}}_R^{-1}\mathbf{J}^{-1}(\hat{\mathbf{r}}^-)\left(\bar{\mathbf{r}} - \hat{\mathbf{r}}^-\right). \tag{205}$$

As with (194), this update also determines a relation between weighted differentials,

$$\hat{\mathbf{X}}^{-1}\left(\hat{\mathbf{x}}^{(m)} - \hat{\mathbf{x}}^-\right) = \bar{\mathbf{X}}_R^{-1}\mathbf{J}^{-1}(\hat{\mathbf{r}}^-)\left(\bar{\mathbf{r}} - \hat{\mathbf{r}}^-\right). \tag{206}$$

Letting $\hat{w}^- = \mathrm{tr}(\hat{\mathbf{X}}^-)^{-1}$ and $\bar{w} \equiv \mathrm{tr}\,\bar{\mathbf{X}}_R^{-1}$, use the second expression in (185) to define

$$\hat{w} \equiv \mathrm{tr}\,\hat{\mathbf{X}}^{-1} = \mathrm{tr}(\hat{\mathbf{X}}^-)^{-1} + \mathrm{tr}\,\bar{\mathbf{X}}_R^{-1}. \tag{207}$$

That is, $\hat{w} = \hat{w}^- + \bar{w}$. And then rewrite (205) as

$$\hat{\mathbf{x}}^{(m)} = \hat{\mathbf{x}}^- + \mathbf{G}\frac{\bar{w}}{\hat{w}}\mathbf{J}^{-1}(\hat{\mathbf{r}}^-)\left(\bar{\mathbf{r}} - \hat{\mathbf{r}}^-\right), \tag{208}$$

where

$$\mathbf{G} \equiv (\hat{w}/\overline{w})\hat{\mathbf{X}}\overline{\mathbf{X}}_{R}^{-1}. \tag{209}$$

In which case, using (197),

$$\mathbf{G}^{-1}\left(\hat{\mathbf{x}}^{(m)} - \hat{\mathbf{x}}^{-}\right) = \frac{r_{\Delta}}{\hat{r}^{-}}\begin{bmatrix} \hat{x}^{-} \\ \hat{y}^{-} \end{bmatrix} + a_{\Delta}\begin{bmatrix} -\hat{y}^{-} \\ +\hat{x}^{-} \end{bmatrix}. \tag{210}$$

Finally, comparing (196) and (210),

$$\mathbf{G}^{-1}\left(\hat{\mathbf{x}}^{(m)} - \hat{\mathbf{x}}^{-}\right) = \hat{\mathbf{x}}^{(s)} - \hat{\mathbf{x}}^{-}. \tag{211}$$

But $\mathbf{G}$ is independent of $\overline{\mathbf{r}}$. Thus, in both the scalar-weight and matrix-weight cases the EKF update introduces certain linearization errors that are not explicitly modeled by (190). When $a_{\Delta} = 0$ they are zero. But when $a_{\Delta} \neq 0$, the unmodeled range error is approximately $|a_{\Delta}\hat{r}^{-}|$ and the unmodeled azimuth error is approximately $a_{\Delta}^{3}/3$. However, as seen in the first expression of (203), the range linearization errors are systematic (they are all positive). In contrast, the sign of the azimuth linearization errors is random – the expected value of their underlying random variable is zero.

5.2 The Preferred Ordering Theorem for the EKF

The findings of the last section lead directly to the Preferred Ordering Theorem (POT): given independent range and azimuth measurements, when using the EKF to update a radar track in rectangular coordinates, the measurements should be used *recursively* in the order azimuth first and range last [2, 3]. Such is outlined in the next two sections, and there some more notation is defined.

5.2.1 The POT in the Scalar-Weight EKF Case

Consider the EKF update defined by (195), but written as

$$\hat{\mathbf{x}}_{\mathbf{r}}^{(s)} = \hat{\mathbf{x}}^{-} + \Delta_{\mathbf{r}}^{(-)}(\overline{\mathbf{r}}) \tag{212}$$

with

$$\Delta_{\mathbf{r}}^{(-)}(\overline{\mathbf{r}}) = \frac{\overline{w}}{\hat{w}} \begin{bmatrix} \cos\hat{a}^- & -\hat{r}^- \sin\hat{a}^- \\ +\sin\hat{a}^- & \hat{r}^- \cos\hat{a}^- \end{bmatrix} \begin{bmatrix} \overline{r} - \hat{r}^- \\ \overline{a} - \hat{a}^- \end{bmatrix}. \tag{213}$$

Here the subscript "$\mathbf{r}$" is used to indicate that the components of the measurement vector are being used together – concurrently, not sequentially – and so (212) with (213) shall be dubbed a *vector*-update. Of course, (212) and (213) may also be written together as

$$\hat{\mathbf{x}}_{\mathbf{r}}^{(s)} = \hat{\mathbf{x}}^- + \Delta_r^{(-)}(\overline{r}) + \Delta_a^{(-)}(\overline{a}), \tag{214}$$

where

$$\Delta_r^{(-)}(\overline{r}) = \frac{\overline{w}}{\hat{w}} \begin{bmatrix} \cos\hat{a}^- \\ +\sin\hat{a}^- \end{bmatrix} \left(\overline{r} - \hat{r}^-\right) \text{ and } \Delta_a^{(-)}(\overline{a}) = \frac{\overline{w}}{\hat{w}} \begin{bmatrix} -\hat{r}^- \sin\hat{a}^- \\ \hat{r}^- \cos\hat{a}^- \end{bmatrix} \left(\overline{a} - \hat{a}^-\right). \tag{215}$$

Here the subscripts "r" and "a" are used to indicate that the scalar measurement components are being used individually.

Next, define two *scalar*-updates as follows: either use $(\overline{r};1/\overline{w})$ first to update $\hat{\mathbf{x}}^-$, or use $(\overline{a};1/\overline{w})$ first to update $\hat{\mathbf{x}}^-$. That is, determine either

$$\hat{\mathbf{x}}_r^{(s)} = \hat{\mathbf{x}}^- + \Delta_r^{(-)}(\overline{r}) \text{ or } \hat{\mathbf{x}}_a^{(s)} = \hat{\mathbf{x}}^- + \Delta_a^{(-)}(\overline{a}). \tag{216}$$

And then use the other measurement component, respectively as

$$\hat{\mathbf{x}}_{\mathbf{r}}^{(s)} = \hat{\mathbf{x}}_r^{(s)} + \Delta_a^{(-)}(\overline{a}) \text{ and } \hat{\mathbf{x}}_{\mathbf{r}}^{(s)} = \hat{\mathbf{x}}_a^{(s)} + \Delta_r^{(-)}(\overline{r}). \tag{217}$$

These are the two sequential scalar-updates corresponding to the above vector update: (217) with (216) provides the same results as (214). If the above scalar-updates were used *recursively*, however, the results would have been different [1, 31].

Define the *range-first azimuth-last recursive update* as follows. First use $(\overline{r};1/\overline{w})$ with the first expression in (216), that is, determine $\hat{\mathbf{x}}_r^{(s)T} = (\hat{x}_r^{(s)}, \hat{y}_r^{(s)})$, which yields $\hat{r}_r^{(s)}$ and $\hat{a}_r^{(s)}$. And then use $(\overline{a};1/\overline{w})$ with $\hat{r}_r^{(s)}$ and $\hat{a}_r^{(s)}$ in the second expression of (215),

$$\Delta_{ra}^{(s)}(\overline{a}) = \frac{\overline{w}}{\hat{w}} \begin{bmatrix} -\hat{r}_r^{(s)} \sin\hat{a}_r^{(s)} \\ \hat{r}_r^{(s)} \cos\hat{a}_r^{(s)} \end{bmatrix} \left(\overline{a} - \hat{a}_r^{(s)}\right), \tag{218}$$

to update $\hat{\mathbf{x}}_r^{(s)}$ as

$$\hat{\mathbf{x}}_{ra}^{(s)} = \hat{\mathbf{x}}_{r}^{(s)} + \Delta_{ra}^{(s)}(\overline{a}) . \tag{219}$$

This yields $\hat{\mathbf{x}}_{ra}^{(s)T} = (\hat{x}_{ra}^{(s)}, \hat{y}_{ra}^{(s)})$, which determines $\hat{r}_{ra}^{(s)}$ and $\hat{a}_{ra}^{(s)}$. (The subscript "ra" is used to denote these outcomes as recursive *range-first azimuth-last* updates.)

Similarly, define the *azimuth-first range-last recursive update* as follows. First use $(\overline{a};1/\overline{w})$ with the second expression in (216), that is, $\hat{\mathbf{x}}_{a}^{(s)T} = (\hat{x}_{a}^{(s)}, \hat{y}_{a}^{(s)})$, which yields $\hat{r}_{a}^{(s)}$ and $\hat{a}_{a}^{(s)}$. And then use $(\overline{r};1/\overline{w})$ with $\hat{r}_{a}^{(s)}$ and $\hat{a}_{a}^{(s)}$ in the first expression of (215),

$$\Delta_{ar}^{(-)}(\overline{r}) = \frac{\overline{w}}{\hat{w}} \begin{bmatrix} \cos \hat{a}_{a}^{-} \\ +\sin \hat{a}_{a}^{-} \end{bmatrix} \left(\overline{r} - \hat{r}_{a}^{-}\right) \tag{220}$$

to update $\hat{\mathbf{x}}_{a}^{(s)}$ as

$$\hat{\mathbf{x}}_{ar}^{(s)} = \hat{\mathbf{x}}_{a}^{(s)} + \Delta_{ar}^{(s)}(\overline{r}) . \tag{221}$$

This yields $\hat{\mathbf{x}}_{ar}^{(s)T} = (\hat{x}_{ar}^{(s)}, \hat{y}_{ar}^{(s)})$, which in turn provides $\hat{r}_{ar}^{(s)}$ and $\hat{a}_{ar}^{(s)}$. Here the subscript " ar" is used to denote these outcomes as *azimuth-first range-last* recursive updates.

5.2.2 The POT in the Matrix-Weight EKF Case

Here the POT equations for the matrix-weight case are presented – along with some new notation for use in the sequel. Specifically, let

$$\mathbf{e}_{\parallel}(a) \equiv \begin{bmatrix} \cos a \\ \sin a \end{bmatrix} \quad \text{and} \quad \mathbf{e}_{\perp}(a) \equiv \begin{bmatrix} +\cos a \\ -\sin a \end{bmatrix}, \tag{222}$$

and write the Jacobian matrices that were defined in Chapter 2 as

$$\mathbf{J}(\boldsymbol{\rho}) = \begin{bmatrix} \mathbf{e}_{\parallel}^{T}(a) \\ r^{-1}\mathbf{e}_{\perp}^{T}(a) \end{bmatrix} \quad \text{and} \quad \mathbf{J}_{\mathbf{r}}^{-1}(\boldsymbol{\rho}) = \left[\mathbf{e}_{\parallel}(a) \quad r\mathbf{e}_{\perp}(a)\right]. \tag{223}$$

And so write the matrix-weight EKF vector-update of $\hat{\mathbf{x}}^{-}$ as

$$\hat{\mathbf{x}}_{\mathbf{r}} = \hat{\mathbf{x}}^{-} + \hat{\mathbf{X}}_{\mathbf{r}}\mathbf{e}_{\parallel}(\hat{a}^{-})\left(\frac{\overline{r} - \hat{r}^{-}}{\sigma_{R}^{2}}\right) + \hat{r}^{-}\hat{\mathbf{X}}_{\mathbf{r}}\mathbf{e}_{\perp}(\hat{a}^{-})\left(\frac{\overline{a} - \hat{a}^{-}}{\sigma_{A}^{2}}\right). \tag{224}$$

(For convenience, the superscript "(m)" is being dropped here to simplify this notation.)

Note that the differential form of the above expression is

$$\hat{\mathbf{X}}_{\mathbf{r}}^{-1}(\hat{\mathbf{x}}_{\mathbf{r}}-\hat{\mathbf{x}}^{-})=\mathbf{e}_{\parallel}(\hat{a}^{-})\left(\frac{\overline{r}-\hat{r}^{-}}{\sigma_R^2}\right)+\hat{r}^{-}\mathbf{e}_{\perp}(\hat{a}^{-})\left(\frac{\overline{a}-\hat{a}^{-}}{\sigma_A^2}\right). \tag{225}$$

In the *sequential* matrix-update case, given $(\hat{\mathbf{x}}^{-};\hat{\mathbf{X}}^{-})$, either use $(\overline{r};\sigma_R^2)$ first,

$$\hat{\mathbf{x}}_r=\hat{\mathbf{x}}^{-}+\hat{\mathbf{X}}_r\mathbf{e}_{\parallel}(\hat{a}^{-})\left(\frac{\overline{r}-\hat{r}^{-}}{\sigma_R^2}\right) \text{ and } \hat{\mathbf{X}}_r=\left[(\hat{\mathbf{X}}^{-})^{-1}+\frac{1}{\sigma_R^2}\mathbf{e}_{\parallel}(\hat{a}^{-})\mathbf{e}_{\parallel}^T(\hat{a}^{-})\right]^{-1}. \tag{226}$$

Or, given $(\hat{\mathbf{x}}^{-};\hat{\mathbf{X}}^{-})$, use $(\overline{a};\sigma_A^2)$ first,

$$\hat{\mathbf{x}}_a=\hat{\mathbf{x}}^{-}+\hat{r}^{-}\hat{\mathbf{X}}_a\mathbf{e}_{\perp}(\hat{a}^{-})\left(\frac{\overline{a}-\hat{a}^{-}}{\sigma_A^2}\right) \text{ and } \hat{\mathbf{X}}_a=\left[(\hat{\mathbf{X}}^{-})^{-1}+\frac{(\hat{r}^{-})^2}{\sigma_A^2}\mathbf{e}_{\perp}(\hat{a}^{-})\mathbf{e}_{\perp}^T(\hat{a}^{-})\right]^{-1}. \tag{227}$$

These two updates respectively imply

$$\hat{\mathbf{X}}_r^{-1}(\hat{\mathbf{x}}_r-\hat{\mathbf{x}}^{-})=\mathbf{e}_{\parallel}(\hat{a}^{-})\left(\frac{\overline{r}-\hat{r}^{-}}{\sigma_R^2}\right) \text{ and } \hat{\mathbf{X}}_a^{-1}(\hat{\mathbf{x}}_a-\hat{\mathbf{x}}^{-})=\hat{r}^{-}\mathbf{e}_{\perp}(\hat{a}^{-})\left(\frac{\overline{a}-\hat{a}^{-}}{\sigma_A^2}\right). \tag{228}$$

And the respective sums of these weighted differentials equals the one in (225). That is,

$$\hat{\mathbf{X}}_{\mathbf{r}}^{-1}\left(\hat{\mathbf{x}}_{\mathbf{r}}-\hat{\mathbf{x}}^{-}\right)=\hat{\mathbf{X}}_r^{-1}(\hat{\mathbf{x}}_r-\hat{\mathbf{x}}^{-})+\hat{\mathbf{X}}_a^{-1}(\hat{\mathbf{x}}_a-\hat{\mathbf{x}}^{-}). \tag{229}$$

Thus,

$$\hat{\mathbf{x}}_{\mathbf{r}}=\hat{\mathbf{x}}^{-}+\mathbf{K}_r\left(\hat{\mathbf{x}}_r-\hat{\mathbf{x}}^{-}\right)+\mathbf{K}_a\left(\hat{\mathbf{x}}_a-\hat{\mathbf{x}}^{-}\right), \tag{230}$$

where

$$\mathbf{K}_r=\hat{\mathbf{X}}_{\mathbf{r}}\hat{\mathbf{X}}_r^{-1} \text{ and } \mathbf{K}_a=\hat{\mathbf{X}}_{\mathbf{r}}\hat{\mathbf{X}}_a^{-1}. \tag{231}$$

That is, as in the scalar-weight case, the sequential-scalar update and the vector update provide the same result.

For the *recursive* matrix-update case, given either $(\hat{\mathbf{x}}_r;\hat{\mathbf{X}}_r)$ and $(\hat{\mathbf{x}}_a;\hat{\mathbf{X}}_a)$ as determined above, (226) and (227), respectively use the other measurement component,

$$\hat{\mathbf{x}}_{ra} = \hat{\mathbf{x}}_r + \hat{\mathbf{X}}_{ra}\mathbf{e}_{\parallel}(\hat{a}_r)\left(\frac{\bar{a}-\hat{a}_r}{\sigma_A^2}\right) \text{ and } \hat{\mathbf{x}}_{ar} = \hat{\mathbf{x}}_a + \hat{\mathbf{X}}_{ar}\mathbf{e}_{\perp}(\hat{a}_a)\left(\frac{\bar{a}-\hat{r}_a}{\sigma_R^2}\right), \tag{232}$$

where

$$\hat{\mathbf{X}}_{ra} = \left[\hat{\mathbf{X}}_r^{-1} + \frac{1}{\sigma_A^2}\mathbf{e}_{\parallel}(\hat{a}_r)\mathbf{e}_{\parallel}^T(\hat{a}_r)\right]^{-1} \text{ and } \hat{\mathbf{X}}_{ar} = \left[\hat{\mathbf{X}}_a^{-1} + \frac{\hat{r}_a^2}{\sigma_R^2}\mathbf{e}_{\parallel}(\hat{a}_a)\mathbf{e}_{\parallel}^T(\hat{a}_a)\right]^{-1}. \tag{233}$$

The updates in (226) and (227) shall be written symbolically as

$$(\hat{\mathbf{x}}^-;\hat{\mathbf{X}}^-)\xrightarrow{(\bar{r};\sigma_R^2)}(\hat{\mathbf{x}}_r;\hat{\mathbf{X}}_r) \text{ and } (\hat{\mathbf{x}}^-;\hat{\mathbf{X}}^-)\xrightarrow{(\bar{a};\sigma_A^2)}(\hat{\mathbf{x}}_a;\hat{\mathbf{X}}_a). \tag{234}$$

And the updates in (232) and (233) shall be written symbolically as

$$(\hat{\mathbf{x}}_r;\hat{\mathbf{X}}_r)\xrightarrow{(\bar{a};\sigma_A^2)}(\hat{\mathbf{x}}_{ra};\hat{\mathbf{X}}_{ra}) \text{ and } (\hat{\mathbf{x}}_a;\hat{\mathbf{X}}_a)\xrightarrow{(\bar{r};\sigma_R^2)}(\hat{\mathbf{x}}_{ar};\hat{\mathbf{X}}_{ar}). \tag{235}$$

Accordingly, these two EKF recursive update cases shall be represented by

$$(\hat{\mathbf{x}}^-;\hat{\mathbf{X}}^-)\xrightarrow{(\bar{r};\sigma_R^2)}(\hat{\mathbf{x}}_r;\hat{\mathbf{X}}_r)\xrightarrow{(\bar{a};\sigma_A^2)}(\hat{\mathbf{x}}_{ra};\hat{\mathbf{X}}_{ra}) \tag{236}$$

and

$$(\hat{\mathbf{x}}^-;\hat{\mathbf{X}}^-)\xrightarrow{(\bar{a};\sigma_A^2)}(\hat{\mathbf{x}}_a;\hat{\mathbf{X}}_a)\xrightarrow{(\bar{r};\sigma_R^2)}(\hat{\mathbf{x}}_{ar};\hat{\mathbf{X}}_{ar}). \tag{237}$$

5.2.3 Illustration of the POT

The EKF and the POT are illustrated below using the $\sigma_A = \pi/12$ case of Chapter 3. Figure 5-1 provides the sample means and sample standard deviations of the estimates of x and y – the solid and dashed curves are respectively those of the EKF and the POT. And Figure 5-2 provides the corresponding converted back cases. Note that the POT does not appear to be very effective for this (very) short range case. (In Chapter 6 a longer-range case is illustrated, and there the POT is seen to be very effective.) In these figures, track initialization transients are also apparent.

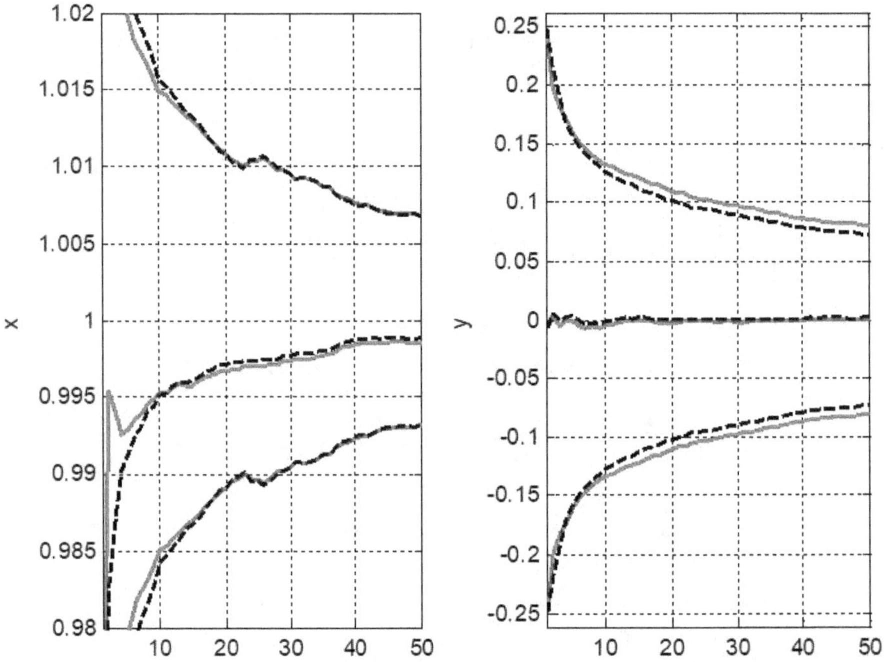

Figure 5-1 Effectiveness of the POT in rectangular coordinates

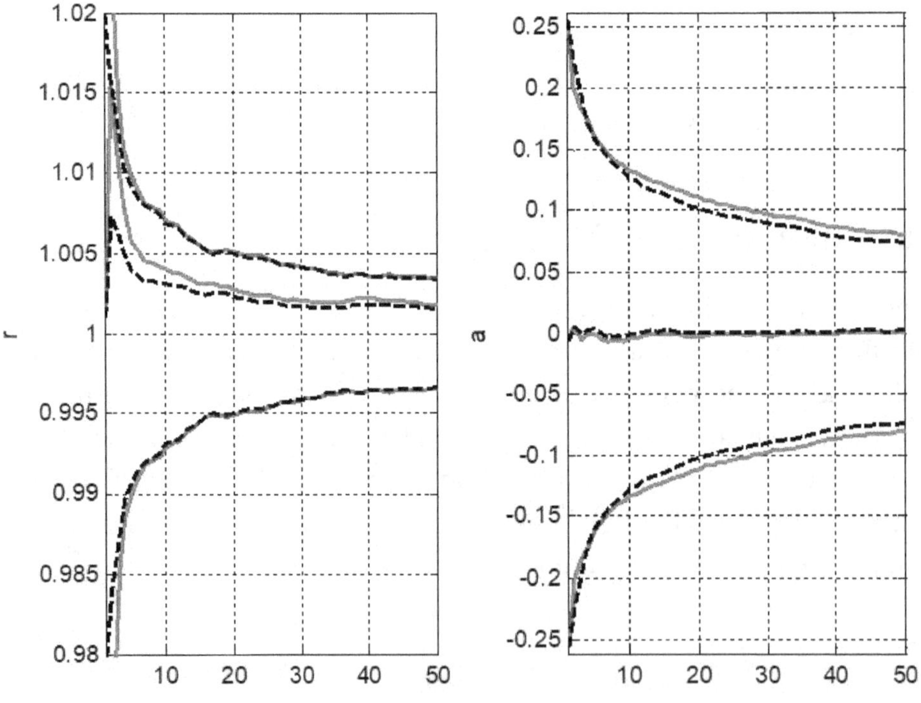

Figure 5-2 Effectiveness of the CB-POT (converted back into radar coordinates)

5.3 The Basic Extended-POT

Given the ineffectiveness of the POT at very short ranges (as illustrated above), the POT shall now be extended as follows – a further extension will be given in the sequel. After updating the estimate using the azimuth measurement pair, $(\bar{a}^-;\sigma_A^2)$, instead of *implicitly* removing just a portion of that spurious linearization error in range (by using the range update after the azimuth update), *explicitly* remove all of it. In particular, after the azimuth-first update, simply restore the estimated range to its prior value; and then update that estimate with $(\bar{r}^-;\sigma_R^2)$.

Let $\hat{\mathbf{x}}_a$ be the outcome of the azimuth-first update defined above. The range it determines is $\hat{r}_a = \sqrt{\hat{x}_a^2 + \hat{y}_a^2}$. But since $\hat{r}^- = \sqrt{(\hat{x}^-)^2 + (\hat{y}^-)^2}$, the prior range corresponding to $\hat{\mathbf{x}}^-$ is known; and the spurious linearization error in range is also known – it is exactly $\hat{r}_a - \hat{r}^-$. Therefore, adjust the estimate after the azimuth-first update, either as

$$\hat{\mathbf{x}}_a^{(*)} \equiv \begin{bmatrix} \hat{r}_a^{(*)} \cos \hat{a}_a^{(*)} \\ \hat{r}_a^{(*)} \sin \hat{a}_a^{(*)} \end{bmatrix} = \begin{bmatrix} \hat{r}_a \cos \hat{a}_a \\ \hat{r}_a \sin \hat{a}_a \end{bmatrix} - \begin{bmatrix} (\hat{r}_a - \hat{r}^-) \cos \hat{a}_a \\ (\hat{r}_a - \hat{r}^-) \sin \hat{a}_a \end{bmatrix} = \begin{bmatrix} \hat{r}^- \cos \hat{a}_a \\ \hat{r}^- \sin \hat{a}_a \end{bmatrix}, \tag{238}$$

or as

$$\hat{\mathbf{x}}_a^{(*)} \equiv \begin{bmatrix} \hat{r}_a^{(*)} \cos \hat{a}_a^{(*)} \\ \hat{r}_a^{(*)} \sin \hat{a}_a^{(*)} \end{bmatrix} = (\hat{r}^- / \hat{r}_a) \begin{bmatrix} \hat{r}_a \cos \hat{a}_a \\ \hat{r}_a \sin \hat{a}_a \end{bmatrix} = \begin{bmatrix} \hat{r}^- \cos \hat{a}_a \\ \hat{r}^- \sin \hat{a}_a \end{bmatrix}. \tag{239}$$

And then use $(\bar{r}^-;\sigma_R^2)$ to update that adjusted estimate instead. That is, determine

$$\hat{\mathbf{x}}_{ar}^{(*)} = \hat{\mathbf{x}}_a^{(*)} + \hat{\mathbf{X}}_{ar}^{(*)} \mathbf{e}_\perp(\hat{a}_a) \left(\frac{\bar{r} - \hat{r}^-}{\sigma_R^2} \right) \quad \text{and} \quad \hat{\mathbf{X}}_{ar}^{(*)} = \left[\hat{\mathbf{X}}_a^{-1} + \frac{(\hat{r}^-)^2}{\sigma_R^2} \mathbf{e}_{\|}(\hat{a}_a) \mathbf{e}_{\|}^T(\hat{a}_a) \right]^{-1}. \tag{240}$$

Figure 5-3 illustrates the effectiveness of this version of the POT (the solid curves), now dubbed the (basic) Extended-POT (EPOT), along with the POT results of Figure 5-1 (the dashed curves). And Figure 5-4 provides the corresponding converted-back cases.

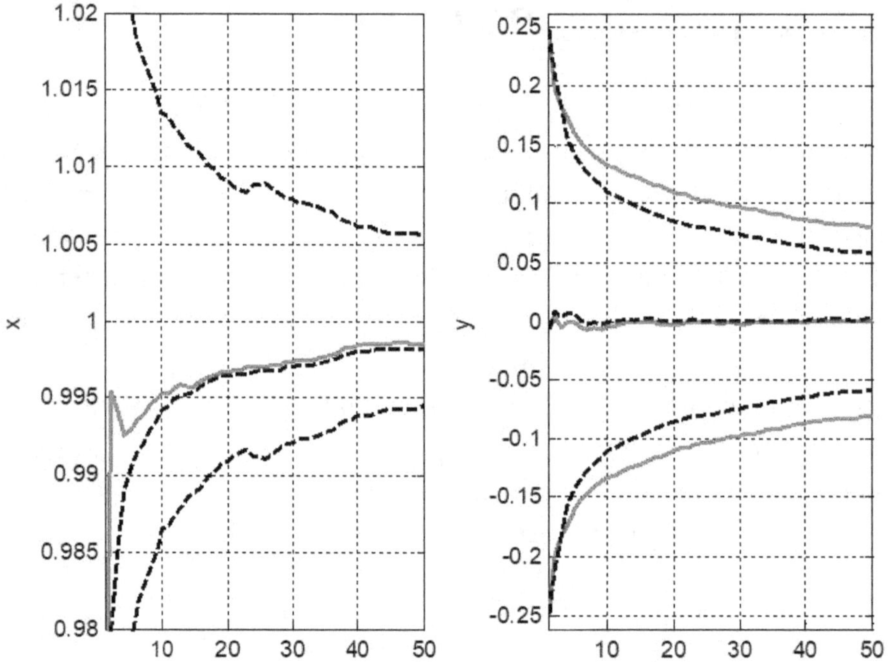

Figure 5-3 Effectiveness of the EPOT in rectangular coordinates

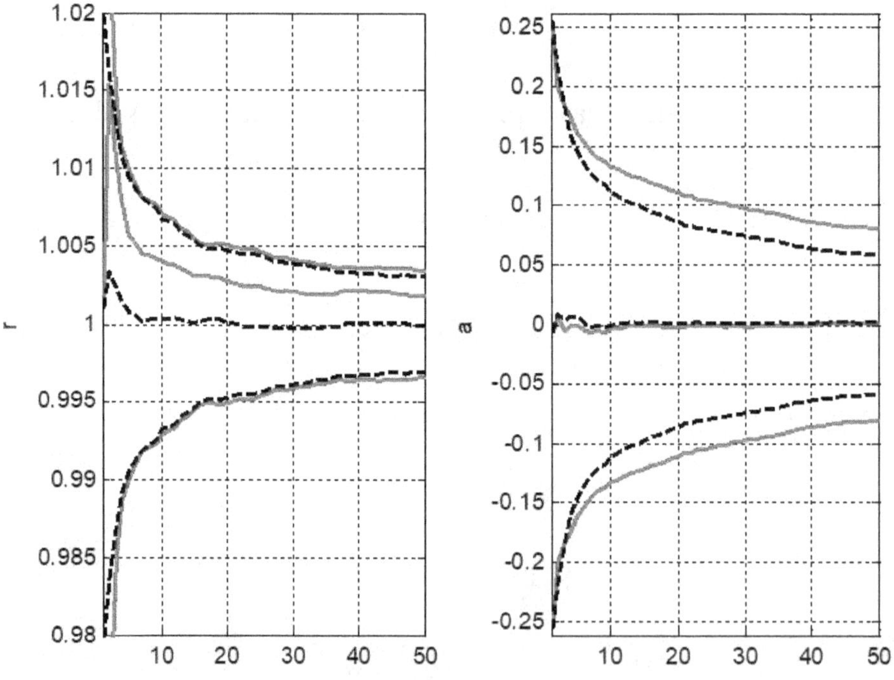

Figure 5-4 Effectiveness of the CB-EPOT (converted back into radar coordinates)

5.4 Chapter 5 References

[1] T. W. Anderson, An Introduction to Multivariate Statistical Analysis, Wiley (2003).

[2] K. S. Miller and D. M. Leskiw, "Nonlinear Estimation with Radar Observations," in Transitions on Aerospace and Electronic Systems, IEEE AES-2, pp. 192-200 (1982).

[3] D. Leskiw and C. Pederson, Metric Accuracy Analysis, Riverside Research Institute Technical Report ESD-TR-79-332 (1978).

6

Application to Radar Tracking

The results of the analyses presented in Chapters 5 and 6 will now be applied to a more stressing tracking problem, the one that has been used in the DCCM/UCCM literature over the years to show the superiority of those "debiasing" methods [1, 2].

The DCCM/UCCM "exemplar" is summarized as follows. The object is initially at 70 kilometers range and 45 degrees azimuth, and its motion has constant velocity, 15 meters per second due north (parallel to the Y-axis). The radar is fixed at the coordinate origin and provides 50 detections, which are at 60 second intervals. The measurements are all unbiased and mutually independent and have gaussian distributions. The range standard deviation is $\sigma_R = 50$ meters. Azimuth has two cases: here $\sigma_A = 1.5$ degrees shall be used (2.5 will be used in the next Chapter). Finally, in that literature the system noise is 0.01 meters/seconds-squared in each coordinate, but here it is zero in all cases.

We will first extend the basic CV tracking equations derived in Chapter 2 to the two degree-of-freedom case, with everything in rectangular coordinates. Then the various forms of the estimators for the CV case are given and illustrated. It will be seen that the EKF with the POT is much better than the "debiased" and "consistent" PLKF. But it shall also be seen that the EPOT as defined in the previous Chapter is not as effective. Here that version is now dubbed the *basic*-EPOT (B-EPOT) – an additional extension to it will be provided in the next Chapter.

6.1 The Basic 2DOF CV Tracking Equations

Let the state vector be $\mathbf{x}^T = (x,\ y,\ \dot{x},\ \dot{y})$. Such shall also be written as $\mathbf{x}^T = (\xi^T,\ \dot{\xi}^T)$, with $\xi^T = (x,\ y)$ and $\dot{\xi}^T = (\dot{x},\ \dot{y})$. Given a prior track at t_{n-1}, $(\hat{\mathbf{x}}_{n-1};\hat{\mathbf{X}}_{n-1})$, the predicted track at t_n is

$$\hat{\mathbf{x}}_n^- = \Phi(\tau)\hat{\mathbf{x}}_{n-1} \text{ and } \hat{\mathbf{X}}_n^- = \Phi(\tau)\hat{\mathbf{X}}_{n-1}\Phi^T(\tau) + \mathbf{S}(\tau), \tag{241}$$

$\tau = t_n - t_{n-1} \geq 0$. The transition and system noise matrices are

$$\Phi(\tau) \equiv \mathbf{I} \otimes \begin{bmatrix} 1 & \tau \\ 0 & 1 \end{bmatrix} = \begin{bmatrix} \mathbf{I} & \tau\mathbf{I} \\ \mathbf{0} & \mathbf{I} \end{bmatrix} \text{ and } \mathbf{S}(\tau) \equiv s\mathbf{I} \otimes \begin{bmatrix} \tau^3/3 & \tau^2/2 \\ \tau^2/2 & \tau \end{bmatrix} \tag{242}$$

(here $s = 0$). Then, given a "rectangular detection" at t_n, $(\overline{\xi}_n;\Sigma_\Xi)$, with $\overline{\xi}_n = \mathbf{H}\mathbf{x}(t_n) + \tilde{\xi}_n$ and $\mathbf{H} = [\mathbf{I} \quad \mathbf{0}] \equiv \mathbf{I} \otimes (1,\ 0)$, the update is

1. $$\hat{\mathbf{x}}_n = \hat{\mathbf{x}}_n^- + \mathbf{K}_n\left(\overline{\xi}_n - \mathbf{H}\hat{\mathbf{x}}_n^-\right) \text{ and } \hat{\mathbf{X}}_n = \left(\mathbf{I} - \mathbf{K}_n\mathbf{H}\right)\hat{\mathbf{X}}_n^-, \tag{243}$$

where

$$\mathbf{K}_n = \hat{\mathbf{X}}_n^-\mathbf{H}^T\left(\mathbf{H}\hat{\mathbf{X}}_n^-\mathbf{H}^T + \Sigma_\Xi\right)^{-1}. \tag{244}$$

The fusion form of the update is

$$\hat{\mathbf{x}}_n = \hat{\mathbf{X}}_n\left[(\hat{\mathbf{X}}_n^-)^{-1}\hat{\mathbf{x}}_n^- + \mathbf{H}^T\Sigma^{-1}\overline{\xi}_n\right] \text{ and } \hat{\mathbf{X}}_n = \left[(\hat{\mathbf{X}}_{n-1}^-)^{-1} + \mathbf{H}^T\Sigma_\Xi^{-1}\mathbf{H}\right]^{-1}. \tag{245}$$

And, using $\mathbf{K}_n = \hat{\mathbf{X}}_n\mathbf{H}^T\Sigma_\Xi^{-1}$, the alternate form of the state vector update equation is

$$\hat{\mathbf{x}}_n = \hat{\mathbf{x}}_n^- + \hat{\mathbf{X}}_n\mathbf{H}^T\Sigma_\Xi^{-1}\left(\overline{\xi}_n - \mathbf{H}\hat{\mathbf{x}}_n^-\right). \tag{246}$$

Now to initialize a CV track at least two detections at distinct times are needed. Here such are said to occur at t_U and t_V, with $t_V > t_U$ (an initially unassociated one followed by an associated one, which *verifies* that the object exists). In particular, given $(\overline{\xi}_U;\Sigma_\Xi)$ and $(\overline{\xi}_V;\Sigma_\Xi)$, with $\tau = t_V - t_U > 0$, use the fusion form to determine

$$\hat{\mathbf{x}}_V \equiv \hat{\mathbf{X}}_V\left[\mathbf{G}_{-\tau}^T\Sigma_\Xi^{-1}\overline{\xi}_U + \mathbf{G}_0^T\Sigma_\Xi^{-1}\overline{\xi}_V\right] \text{ and } \hat{\mathbf{X}}_V^{-1} \equiv \mathbf{G}_{-\tau}^T\Sigma_\Xi^{-1}\mathbf{G}_{-\tau} + \mathbf{G}_0^T\Sigma_\Xi^{-1}\mathbf{G}_0, \tag{247}$$

with $\mathbf{G}_\tau \equiv \mathbf{H}\Phi(\tau)$. Note that

$$\mathbf{G}_{\pm\tau} = \begin{bmatrix} 1 & 0 & \pm\tau & 0 \\ 0 & 1 & 0 & \pm\tau \end{bmatrix} \equiv \mathbf{I} \otimes (1, \ \pm\tau). \tag{248}$$

Also, $\mathbf{G}_0 = \mathbf{H}$.

Now

$$\mathbf{G}_{\pm\tau}^T \boldsymbol{\Sigma}^{-1} \mathbf{G}_{\pm\tau} = \begin{bmatrix} \mathbf{I} \\ \pm\tau\mathbf{I} \end{bmatrix} \boldsymbol{\Sigma}_{\Xi}^{-1} \begin{bmatrix} \mathbf{I} & \pm\tau\mathbf{I} \end{bmatrix} \equiv \boldsymbol{\Sigma}_{\Xi}^{-1} \otimes \begin{bmatrix} 1 & \pm\tau \\ \pm\tau & \tau^2 \end{bmatrix}. \tag{249}$$

And so

$$\hat{\mathbf{x}}_V = \hat{\mathbf{X}}_V \begin{bmatrix} \boldsymbol{\Sigma}_{\Xi}^{-1}\bar{\xi}_U + \boldsymbol{\Sigma}_{\Xi}^{-1}\bar{\xi}_V \\ -\tau\boldsymbol{\Sigma}_{\Xi}^{-1}\bar{\xi}_U \end{bmatrix} \text{ and } \hat{\mathbf{X}}_V^{-1} = \begin{bmatrix} 2\boldsymbol{\Sigma}_{\Xi}^{-1} & -\tau\boldsymbol{\Sigma}_{\Xi}^{-1} \\ -\tau\boldsymbol{\Sigma}_{\Xi}^{-1} & \tau^2\boldsymbol{\Sigma}_{\Xi}^{-1} \end{bmatrix}. \tag{250}$$

Thus,

$$\hat{\mathbf{x}}_V = \begin{bmatrix} \hat{\xi}_V \\ \hat{\dot{\xi}}_V \end{bmatrix} = \begin{bmatrix} \bar{\xi}_V \\ (\bar{\xi}_U - \bar{\xi}_V)/\tau \end{bmatrix} \text{ and } \hat{\mathbf{X}}_V = \boldsymbol{\Sigma}_{\Xi} \otimes \begin{bmatrix} 1 & 1/\tau \\ 1/\tau & 2/\tau^2 \end{bmatrix}. \tag{251}$$

For convenience, write $(\hat{\mathbf{x}}_0;\hat{\mathbf{X}}_0) = (\hat{\mathbf{x}}_V;\hat{\mathbf{X}}_V)$ and $t_0 = t_V$; and reindex the subsequent detections as $n = 1, 2, \cdots, N$. Then, using the special case given in Chapter 2, for $n = 1, 2, \cdots, N$, the updated "covariance" matrices are

$$\hat{\mathbf{X}}_n = \boldsymbol{\Sigma}_{\Xi} \otimes \begin{bmatrix} 2(2m-1)/m(m+1) & 6/m(m+1)\tau \\ 6/m(m+1)\tau & 12/(m^2-1)m\tau^2 \end{bmatrix}, \tag{252}$$

$m = n+2$. In which case, the corresponding gain matrix is

$$\mathbf{K}_n = \hat{\mathbf{X}}_n \mathbf{H}^T \boldsymbol{\Sigma}_{\Xi}^{-1} = \mathbf{I} \otimes \begin{bmatrix} 2(2m-1)/m(m+1) \\ 6/m(m+1)\tau \end{bmatrix}. \tag{253}$$

6.1.1 The 2DOF CV LKF Case

Now in the CV LKF case the above equations are used formally with everything in radar coordinates. The state vector is $\mathbf{r}^T = (\boldsymbol{\rho}^T, \dot{\boldsymbol{\rho}}^T)$, with $\boldsymbol{\rho}^T = (r, a)$ and $\boldsymbol{\rho}^T = (\dot{r}, \dot{a})$. And the measurement model is $\bar{\boldsymbol{\rho}}_n = \mathbf{H}\mathbf{r}(t_n) + \tilde{\boldsymbol{\rho}}_n$, with $\mathbf{H} = [\mathbf{I} \ \ \mathbf{0}]$. For our initialization here, $(\bar{\boldsymbol{\rho}}_U;\boldsymbol{\Sigma}_R)$ and $(\bar{\boldsymbol{\rho}}_V;\boldsymbol{\Sigma}_R)$ are given, and the initial track is determined as

$$\hat{\mathbf{r}}_0 = \begin{bmatrix} \hat{\boldsymbol{\rho}}_0 \\ \hat{\dot{\boldsymbol{\rho}}}_0 \end{bmatrix} = \begin{bmatrix} \bar{\boldsymbol{\rho}}_V \\ (\bar{\boldsymbol{\rho}}_U - \bar{\boldsymbol{\rho}}_V)/\tau \end{bmatrix} \text{ and } \hat{\mathbf{R}}_0 = \boldsymbol{\Sigma}_R \otimes \begin{bmatrix} 1 & 1/\tau \\ 1/\tau & 2/\tau^2 \end{bmatrix}. \tag{254}$$

Then, given $(\hat{\mathbf{r}}_{n-1};\hat{\mathbf{R}}_{n-1})$, $n = 1,2,\cdots,N$, the predicted LKF track at t_n is

$$\hat{\mathbf{r}}_n^- = \Phi(\tau)\hat{\mathbf{r}}_{n-1} \text{ and } \hat{\mathbf{R}}_n^- = \Phi(\tau)\hat{\mathbf{R}}_{n-1}\Phi^T(\tau) + \mathbf{S}(\tau), \tag{255}$$

with Φ and $\mathbf{S}$ as in (242). Note that "CV" is a misnomer: this model defines a spiral in $\mathbb{E}$. In practice, this model is used for tracking incoming objects that are at very long ranges, and for "tracking" clutter. Given $(\bar{\boldsymbol{\rho}}_n;\boldsymbol{\Sigma}_R)$, the track is updated as

$$\hat{\mathbf{r}}_n = \hat{\mathbf{r}}_n^- + \mathbf{K}_n\left(\bar{\boldsymbol{\rho}}_n - \mathbf{H}\hat{\mathbf{r}}_n^-\right) \text{ and } \hat{\mathbf{R}}_n = \left(\mathbf{I} + \mathbf{K}_n\mathbf{H}_n\right)\hat{\mathbf{R}}_n^-, \tag{256}$$

where

$$\mathbf{K}_n = \hat{\mathbf{R}}_n^-\mathbf{H}^T\left(\mathbf{H}\hat{\mathbf{R}}_n^-\mathbf{H}^T + \boldsymbol{\Sigma}_R\right)^{-1}, \tag{257}$$

with the components of this $\mathbf{K}_n$ the same as those in (253).

Figure 6-1 illustrates this LKF case with 100 Monte Carlo trials. The left-hand side shows the radar measurements plotted in $\mathbb{E}$, where the problem is defined. And the right-hand side shows the LKF estimates of position, again in $\mathbb{E}$. The darker curves down the centers are respectively the sample means of the measurements and estimates. Figure 6-2 provides the position errors of these LKF estimates; and there the dark curves represent the sample means of the respective sets of errors – the top two plots provide the errors in radar coordinates, and the bottom two plots provide the errors in rectangular coordinates. Note that this LKF is biased – because its model of motion defines a spiral. An improvement will be given in the next Section.

6.1.2 The 2DOF CV C-LKF Case

The Radar Principal Cartesian Coordinates (RPCC) method shall now be used to illustrate the C-LKF case [3]. The updates are determined in radar coordinates, and the covariance matrix is propagated in radar coordinates, but the estimated state vector is propagated in rectangular coordinates. For the sake of comparison with the previous results, however, the same LKF initial track used above will be reused here, $(\hat{\mathbf{r}}_0;\hat{\mathbf{R}}_0) \equiv (\hat{\mathbf{r}}_V;\hat{\mathbf{R}}_V)$.

First, to propagate the estimated state vectors in rectangular coordinates, given $\hat{\mathbf{r}}_{n-1}^T = (\hat{\boldsymbol{\rho}}_{n-1}^T, \hat{\dot{\boldsymbol{\rho}}}_{n-1}^T)$, transform the components of $\hat{\boldsymbol{\rho}}_{n-1}^T = (\hat{r}_{n-1}, \hat{a}_{n-1})$ and $\hat{\dot{\boldsymbol{\rho}}}_{n-1}^T = (\hat{\dot{r}}_{n-1}, \hat{\dot{a}}_{n-1})$.

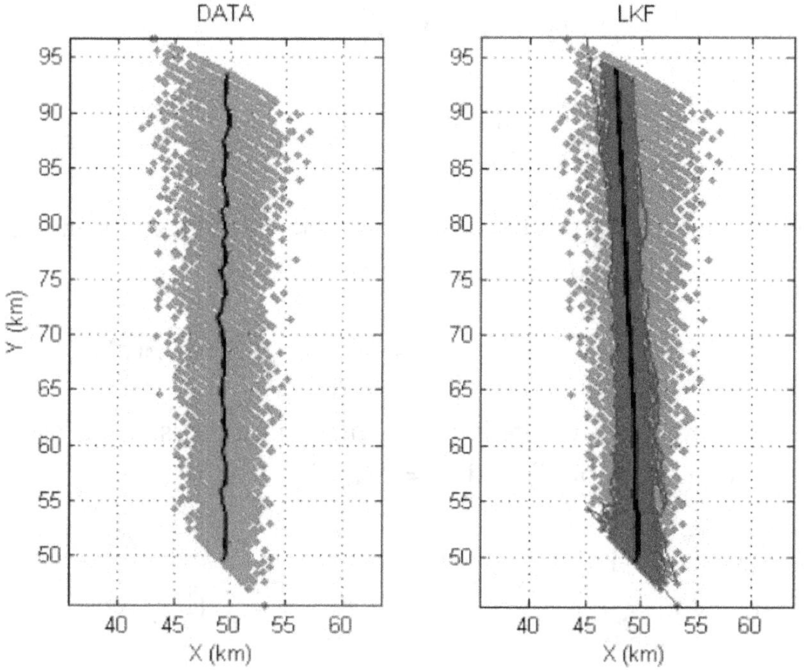

Figure 6-1 Radar measurements and LKF position estimates (in Euclidean space)

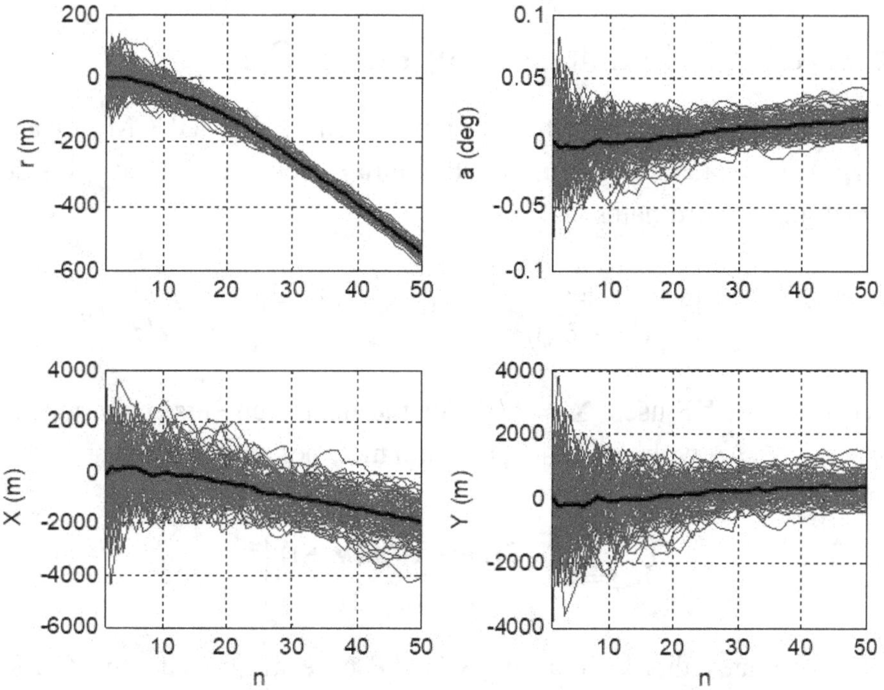

Figure 6-2 LKF position estimation errors (2DOF CV case)

respectively as $\hat{\xi}_{n-1} = \mathbf{h}^{-1}(\hat{\boldsymbol{\rho}}_{n-1})$ and $\hat{\dot{\xi}}_{n-1} = \mathbf{J}^{-1}(\hat{\boldsymbol{\rho}}_{n-1})\hat{\dot{\boldsymbol{\rho}}}_{n-1}$. That is,

$$\xi = \begin{bmatrix} x \\ y \end{bmatrix} = \begin{bmatrix} r\cos a \\ r\sin a \end{bmatrix} \text{ and } \dot{\xi} = \begin{bmatrix} \dot{x} \\ \dot{y} \end{bmatrix} = \begin{bmatrix} \cos a & -r\sin a \\ +\sin a & r\cos a \end{bmatrix} \begin{bmatrix} \dot{r} \\ \dot{a} \end{bmatrix}. \quad (258)$$

And then form $\hat{\mathbf{x}}_{n-1}^T = (\hat{\xi}_{n-1}^T, \hat{\dot{\xi}}_{n-1}^T)$. In which case the initial conditions have the form $(\hat{\mathbf{x}}_{n-1}; \hat{\mathbf{R}}_{n-1})$. And the propagation equations are respectively

$$\hat{\mathbf{x}}_n^- = \Phi(\tau)\hat{\mathbf{x}}_{n-1} \text{ and } \hat{\mathbf{R}}_n^- = \Phi(\tau)\hat{\mathbf{R}}_{n-1}\Phi^T(\tau) + \mathbf{S}(\tau). \quad (259)$$

For the updates, the predicted estimate in rectangular coordinates, $\hat{\mathbf{x}}_n^-$, is transformed back into radar coordinates as $\hat{\boldsymbol{\rho}}_n^- = \mathbf{h}(\hat{\xi}_n^-)$ and $\hat{\dot{\boldsymbol{\rho}}}_n = \mathbf{J}(\hat{\xi}_n)\hat{\dot{\xi}}_n$. That is,

$$\hat{\boldsymbol{\rho}}_n^- = \begin{bmatrix} \sqrt{(\hat{x}_n^-)^2 + (\hat{y}_n^-)^2} \\ \arctan(\hat{y}_n^-, \hat{x}_n^-) \end{bmatrix} \text{ and } \hat{\dot{\boldsymbol{\rho}}}_n^- = \begin{bmatrix} \cos\hat{a}_n^- & \sin\hat{a}_n^- \\ -(1/\hat{r}_n^-)\sin\hat{a}_n^- & (1/\hat{r}_n^-)\cos\hat{a}_n^- \end{bmatrix} \begin{bmatrix} \hat{\dot{x}}_n^- \\ \hat{\dot{y}}_n^- \end{bmatrix}. \quad (260)$$

And $(\hat{\mathbf{r}}_n^-; \hat{\mathbf{R}}_n^-)$, where $(\hat{\mathbf{r}}_n^-)^T = ((\hat{\boldsymbol{\rho}}_n^-)^T, (\hat{\dot{\boldsymbol{\rho}}}_n^-)^T)$, is updated using $(\overline{\boldsymbol{\rho}}_n; \boldsymbol{\Sigma}_{\boldsymbol{R}})$ with (256) and (257). Figure 6-3 illustrates these C-LKF estimates, along with the corresponding LKF ones shown above. Figure 6-4 provide the position errors.

6.1.3 The 2DOF CV Scalar-Weight PLKF Case

The "debiased" S-PLKF is illustrated next, using the UCCM measurements, $\overline{\xi}_n = (1/\lambda)\mathbf{h}^{-1}(\overline{\boldsymbol{\rho}}_n)$, with $\lambda = e^{-\sigma_A^2/2}$. But now the initial track, $(\hat{\mathbf{x}}_0; \hat{\mathbf{X}}_0)$ at t_0, is determined directly in rectangular coordinates, as

$$\hat{\mathbf{x}}_0 = \begin{bmatrix} \hat{\xi}_V \\ \hat{\dot{\xi}}_V \end{bmatrix} = \begin{bmatrix} \overline{\xi}_V \\ (\overline{\xi}_U - \overline{\xi}_V)/\tau \end{bmatrix} \text{ and } \hat{\mathbf{X}}_0 = \overline{\mathbf{X}}_0 \otimes \begin{bmatrix} 1 & 1/\tau \\ 1/\tau & 2/\tau^2 \end{bmatrix}. \quad (261)$$

Recall that the S-PLKF uses $\overline{\mathbf{X}}_n = \mathbf{I}/\overline{w}$ for the measurements' covariance matrices. And so for $n = 1, 2, \cdots, N$, with $m = n + 2$, the gain matrices are the same as in the previous two examples,

$$\mathbf{K}_n = \mathbf{I} \otimes \begin{bmatrix} 2(2m-1)/m(m+1) \\ 6/m(m+1)\tau \end{bmatrix}. \quad (262)$$

Figure 6-5 illustrates this "debiased" S-PLKF case along with the C-LKF results shown above. The estimation position errors are provided by Figure 6-6.

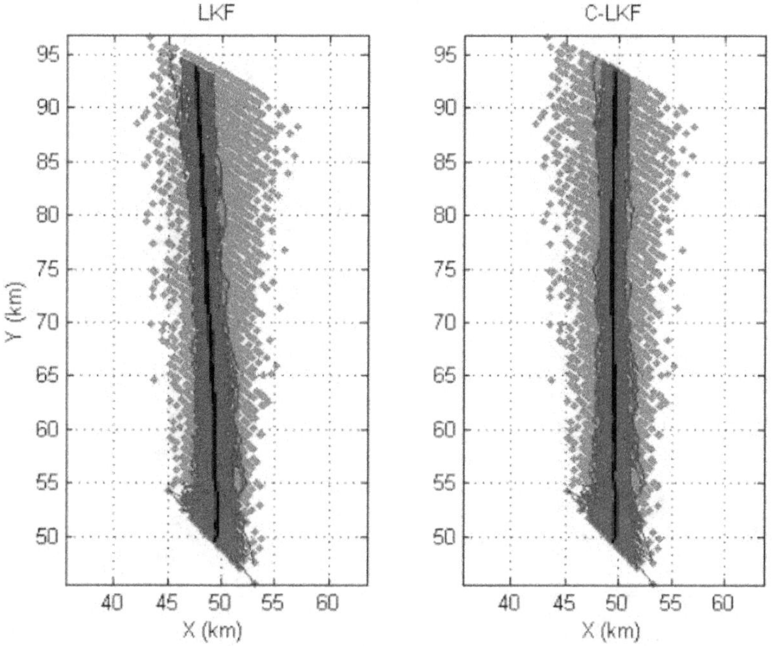

Figure 6-3 Comparison of the LKF and C-LKF (in Euclidean space)

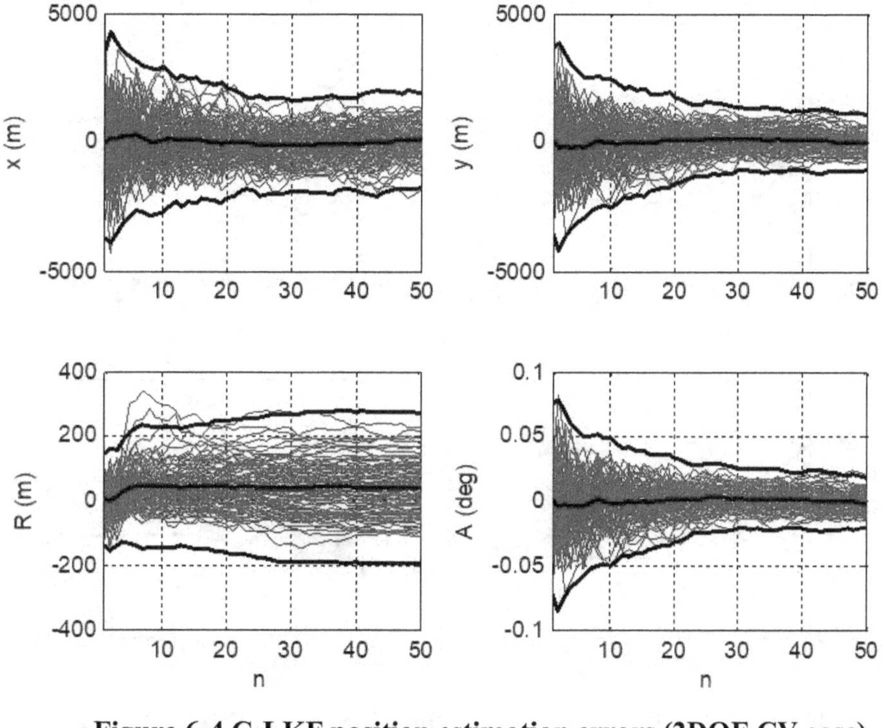

Figure 6-4 C-LKF position estimation errors (2DOF CV case)

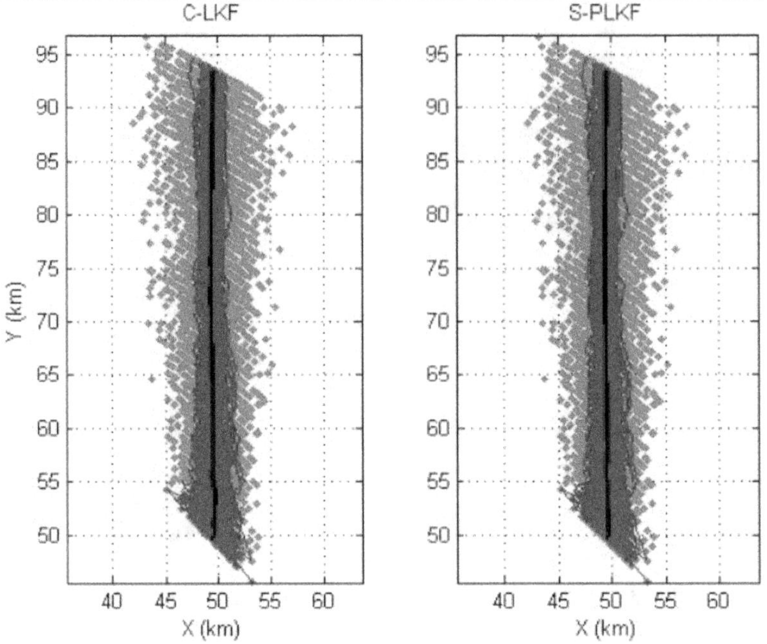

Figure 6-5 Comparison of the C-LKF and S-PLKF (in Euclidean space)

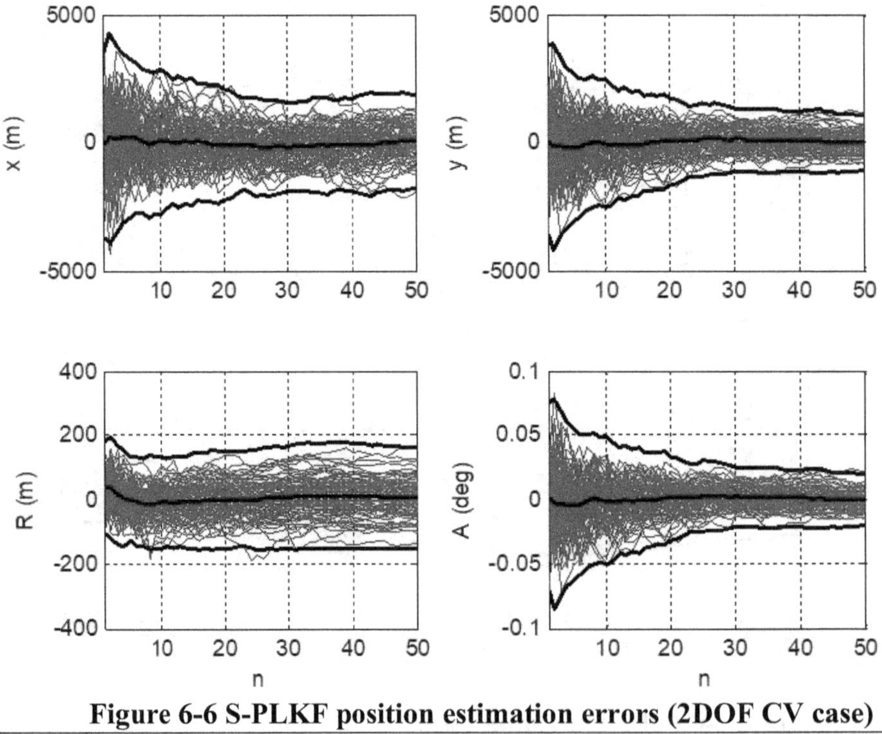

Figure 6-6 S-PLKF position estimation errors (2DOF CV case)

6.1.4 The 2DOF CV Matrix-Weight PLKF Case

Here the *prior-weight* UCCM M-PLKF is illustrated. That is, $\overline{\mathbf{X}}_n = \mathbf{\Sigma}_{X'}\left(\overline{\xi}_{n-1}(\lambda)\right)$, where

$$\mathbf{\Sigma}_{X'}(\mathbf{x}) = \left(\sigma_R^2 + r^2\right)\left\{\frac{1-e^{-2\sigma_A^2}}{2}\begin{bmatrix}1 & 0\\ 0 & 1\end{bmatrix} + \frac{e^{-2\sigma_A^2}}{r^2}\begin{bmatrix}x^2 & xy\\ xy & y^2\end{bmatrix}\right\} - e^{-\sigma_A^2}\begin{bmatrix}x^2 & xy\\ yx & y^2\end{bmatrix}. \quad (263)$$

As before, the propagation equations are

$$\hat{\mathbf{x}}_n^- = \Phi(\tau)\hat{\mathbf{x}}_{n-1} \text{ and } \hat{\mathbf{X}}_n^- = \Phi(\tau)\hat{\mathbf{X}}_{n-1}\Phi^T(\tau) + \mathbf{S}(\tau), \quad (264)$$

and the update equations are

$$\hat{\mathbf{x}}_n = \hat{\mathbf{x}}_n^- + \mathbf{K}_n\left(\overline{\xi}_n - \mathbf{H}\hat{\mathbf{x}}_n^-\right) \text{ and } \hat{\mathbf{X}}_n = \left(\mathbf{I} - \mathbf{K}_n\mathbf{H}_n\right)\hat{\mathbf{X}}_n^-, \quad (265)$$

$$\mathbf{K}_n = \hat{\mathbf{X}}_n^-\mathbf{H}^T\left(\mathbf{H}\hat{\mathbf{X}}_n^-\mathbf{H}^T + \overline{\mathbf{X}}_n\right)^{-1}. \quad (266)$$

Figure 6-7 illustrates the estimates for this "debiased and consistent" M-PLKF, along with those of the "debiased" S-LKF case shown in the previous section. Figure 6-8 provides the corresponding estimation errors.

6.1.5 Discussion of the 2DOV CV C-LKF, S-PLKF, and M-PLKF Cases

Figure 6-9 compares the sample mean-errors of the C-LKF, S-PLKF, and M-PLKF position estimates that were shown above (but here 500 Monte Carlo trials are used). Figure 6-10 provides the corresponding sample standard deviations.

In each figure the top sets of plots are in rectangular coordinates while the lower sets are in radar coordinates (the latter are the "converted-back" cases – those sample means and sample standard deviations being determined after transforming the estimates).

Note that the mean-errors in *range* of the "debiased and consistent" M-LKF appear to be worse than the "debiased" S-PLKF ones. However, for $n > 20$, the sample standard deviations of those M-PLKF estimates are better than the ones of the C-LKF and S-PLKF estimates. That is because the C-LKF and S-PLKF measurement-weights, $\overline{w}$, are independent of range and azimuth, while the M-PLKF ones scale as $1/(\sigma_R^2 + r_n^2\sigma_A^2)$.

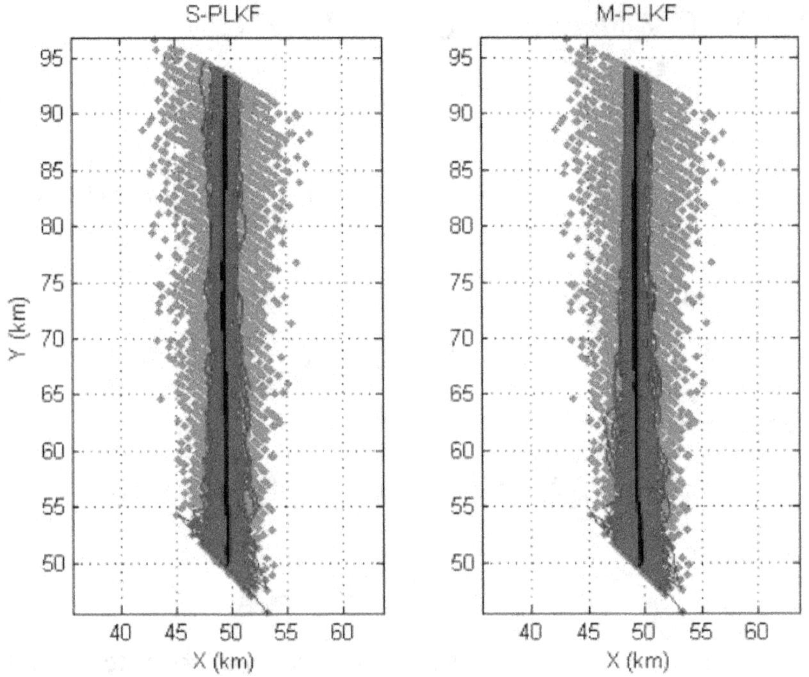

Figure 6-7 Comparison of the S-PLKF and M-PLKF (in Euclidean space)

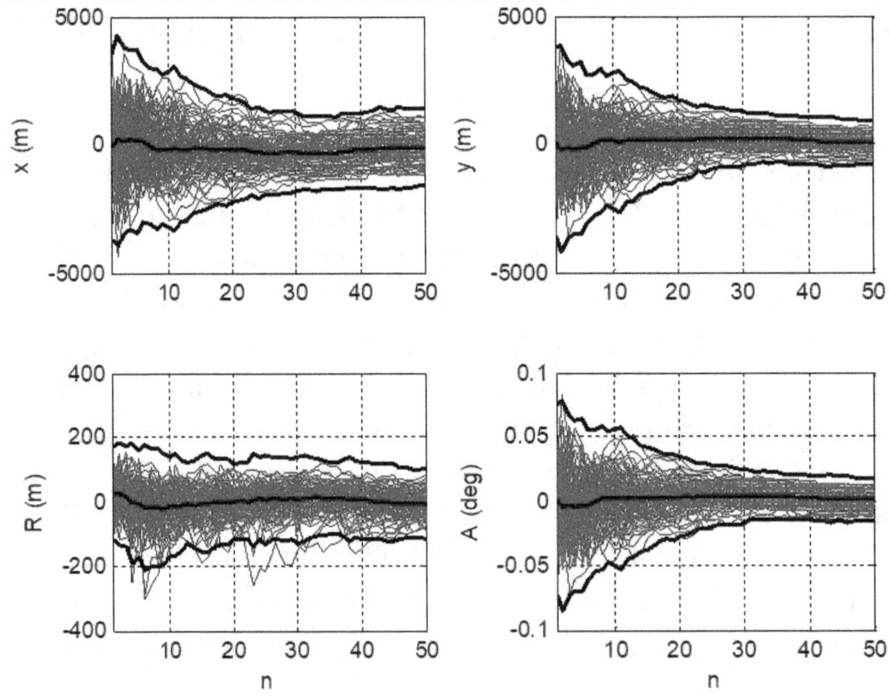

Figure 6-8 M-PLKF position estimation errors (2DOF CV case)

6.2 The 2DOF CV EKF and POT Cases

Here the performance of the EKF is illustrated, first with and then without the POT. For the sake of comparison with the previous results, all the initial tracks are the same as the one used above in the (prior-weight UCCM) M-PLKF case given above. And the propagation equations are also the same as in that M-PLKF case.

6.2.1 The 2DOF CV EKF without the POT

Recall that the (vector) EKF update equations are

$$\hat{\mathbf{x}} = \hat{\mathbf{x}}^- + \mathbf{K}\left(\overline{\boldsymbol{\rho}} - \mathbf{h}(\hat{\boldsymbol{\xi}}^-)\right) \text{ and } \hat{\mathbf{X}} = \left(\mathbf{I} - \mathbf{K}\mathbf{H}^-\right)\hat{\mathbf{X}}^- \tag{267}$$

and

$$\mathbf{K} = \hat{\mathbf{X}}^-\mathbf{H}^T\left(\mathbf{H}\hat{\mathbf{X}}^-\mathbf{H}^T + \boldsymbol{\Sigma}_R\right)^{-1}. \tag{268}$$

Figure 6-11 illustrates this EKF case with 100 Monte Carlo trials, along with the "debiased and consistent" M-PLKF ones shown earlier. Figure 6-12 provide the errors of these EKF estimates.

6.2.2 The 2DOF CV EKF with the POT

Recall that the POT uses the measurement components of a radar detection recursively, $(\overline{a};\sigma_A^2)$ first and $(\overline{r};\sigma_R^2)$ last. In particular, given $(\hat{\mathbf{x}}^-;\hat{\mathbf{X}}^-)$ and $(\overline{a};\sigma_A^2)$, the azimuth-first scalar-update with $\mathbf{H}_a \equiv da/d\mathbf{x}^T$ is

$$\hat{\mathbf{x}}_a = \hat{\mathbf{x}}^- + \mathbf{K}_a(\overline{a} - \hat{a}^-) \text{ and } \hat{\mathbf{X}}_a = \left(\mathbf{I} - \mathbf{K}_a\mathbf{H}_a(\hat{\mathbf{x}}^-)\right)\hat{\mathbf{X}}^- \tag{269}$$

$$\mathbf{K}_a = \hat{\mathbf{X}}^-\mathbf{H}_a^T(\hat{\mathbf{x}}^-)\left(\mathbf{H}_a(\hat{\mathbf{x}}^-)\hat{\mathbf{X}}^-\mathbf{H}_a^T(\hat{\mathbf{x}}^-) + \sigma_A^2\right)^{-1}. \tag{270}$$

And, given $(\hat{\mathbf{x}}_a;\hat{\mathbf{X}}_a)$ and $(\overline{r};\sigma_R^2)$, the range-last scalar-update with $\mathbf{H}_r \equiv dr/d\mathbf{x}^T$ is

$$\hat{\mathbf{x}}_{ar} = \hat{\mathbf{x}}_a + \mathbf{K}_{ar}(\overline{r} - \hat{r}_a) \text{ and } \hat{\mathbf{X}}_{ar} = \left(\mathbf{I} - \mathbf{K}_{ar}\mathbf{H}_r(\hat{\mathbf{x}}_a)\right)\hat{\mathbf{X}}_a \tag{271}$$

$$\mathbf{K}_{ar} = \hat{\mathbf{X}}_a\mathbf{H}_r^T(\hat{\mathbf{x}}_a)\left(\mathbf{H}_r(\hat{\mathbf{x}}_a)\hat{\mathbf{X}}_a\mathbf{H}_r^T(\hat{\mathbf{x}}_a) + \sigma_R^2\right)^{-1}. \tag{272}$$

Figure 6-13 illustrates this EKF with the POT case, using the same 100 Monte Carlo trial data, along with the basic EKF estimates that were shown above. And Figure 6-14 provides the corresponding position errors.

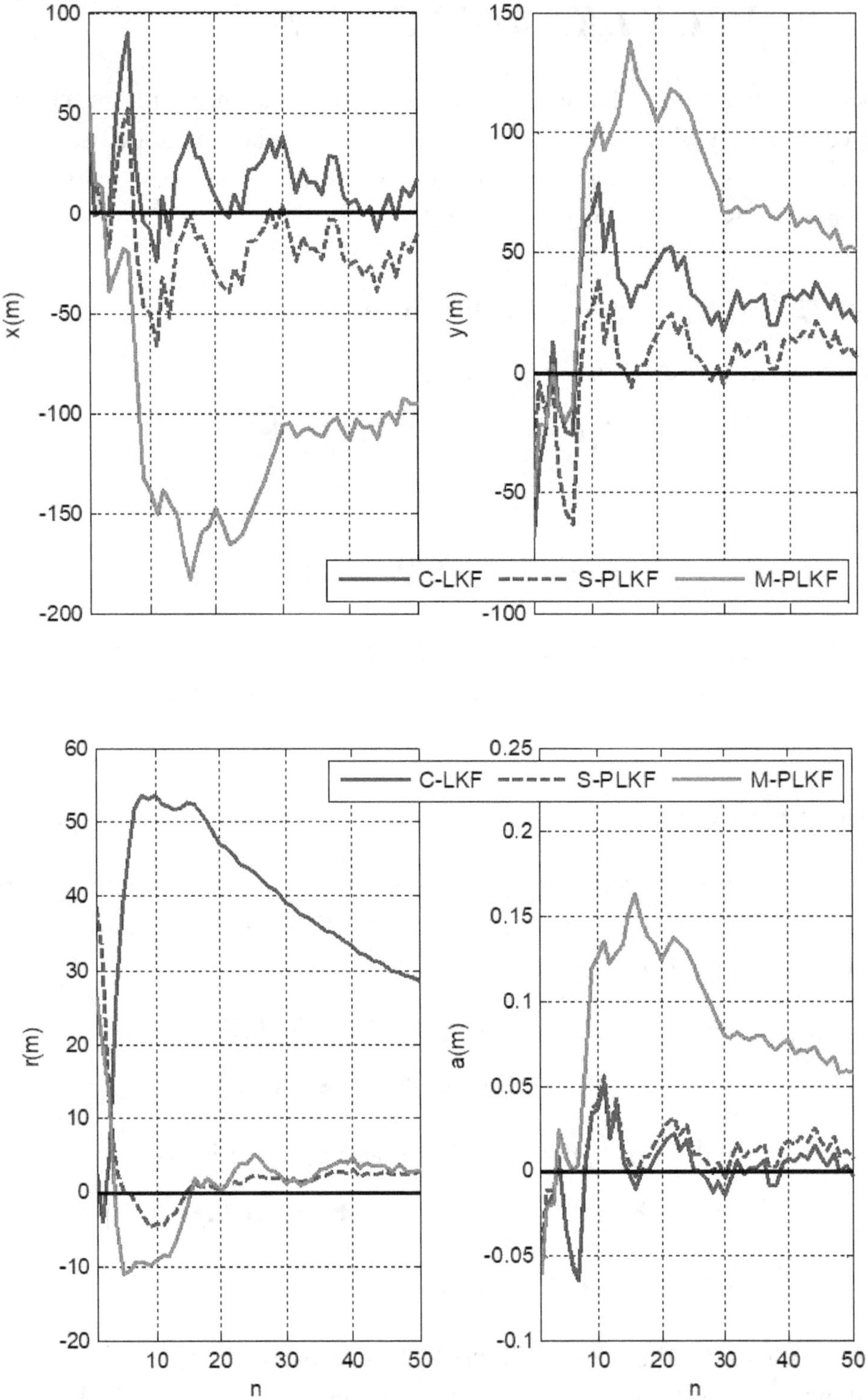

Figure 6-9 Sample means of C-LKF, S-PLKF, and M-PLKF errors (CV 2DOF)

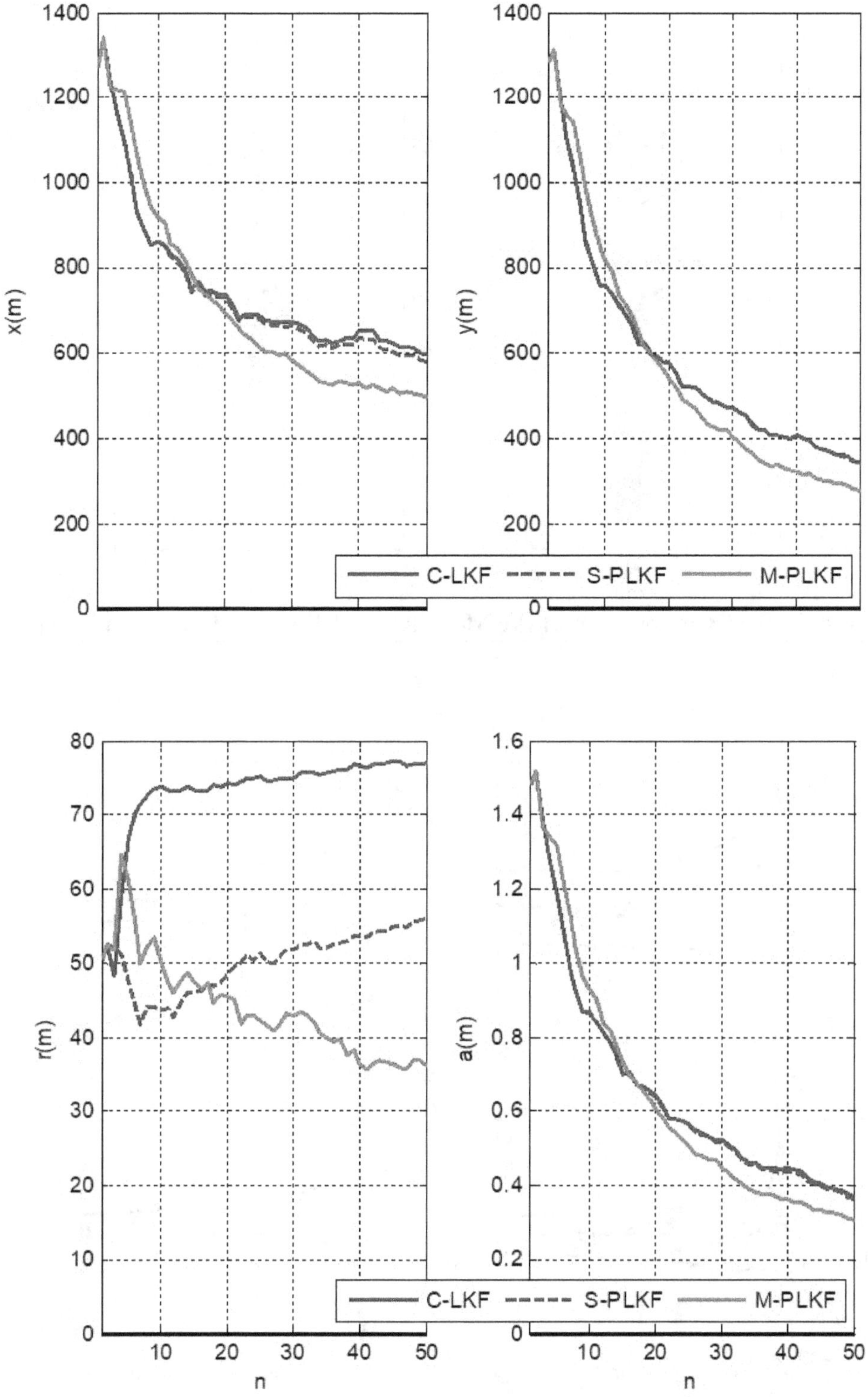

Figure 6-10 Sample standard deviations of C-LKF, S-PLKF, M-PLKF (CV 2DOF)

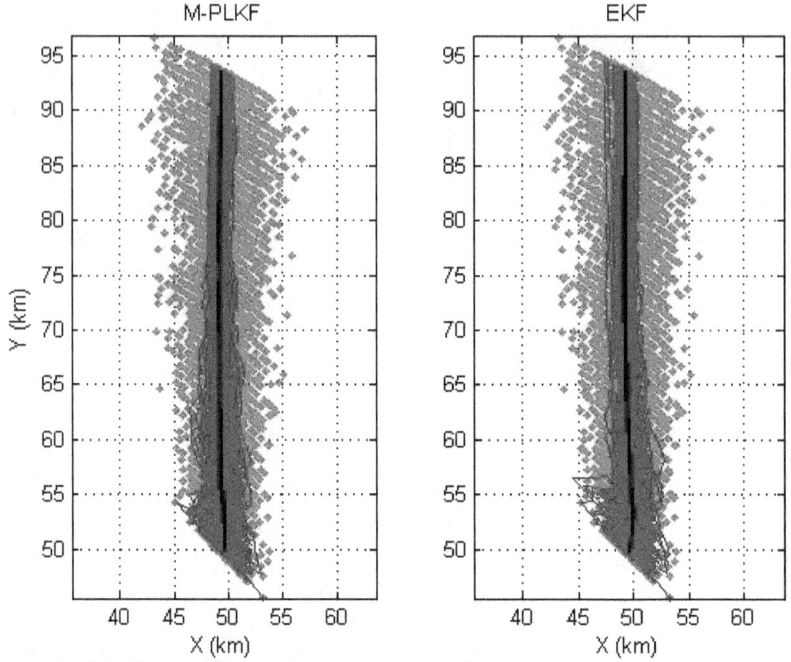

Figure 6-11 Comparison of the M-PLKF and EKF (in Euclidean space)

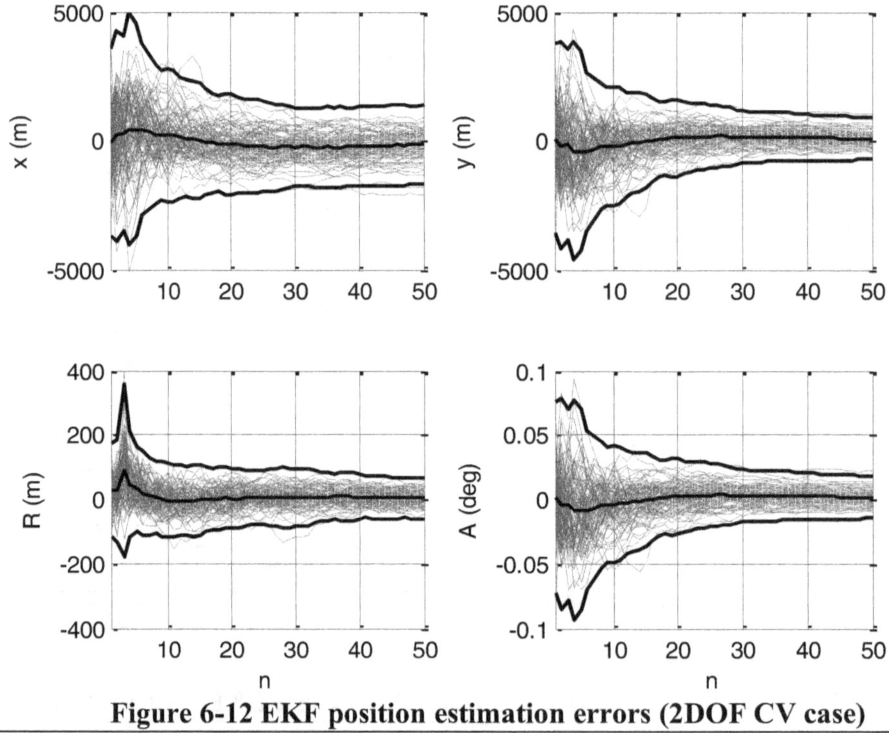

Figure 6-12 EKF position estimation errors (2DOF CV case)

Note that the estimates determined by the EKF with the POT appear to be better than those determined in the previous section by the EKF. Also in the EKF without that POT case, there were a few "estimation paths" that did not seem to converge very well. Here that problem does not seem to appear.

6.3 The 2DOF CV Case with the Basic-EPOT

In the previous Chapter the EPOT "extended" the POT as follows: after the azimuth update, the range of the track is restored to its prior value and the range update is then determined. So given the scalar azimuth-first update, $(\hat{\mathbf{x}}_a;\hat{\mathbf{X}}_a)$, let $\hat{\xi}_a^{(*)} = (r^-/r_a)\hat{\xi}_a$ and form $\hat{\mathbf{x}}_a^{(*)T} = (\hat{\xi}_a^{(*)T},\ \dot{\hat{\xi}}_a^T)$. And then use the range measurement as

$$\hat{\mathbf{x}}_{ar}^{(*)} = \hat{\mathbf{x}}_a^{(*)} + \mathbf{K}_{ar}^{(*)}(\bar{r} - \hat{r}^-) \quad \text{and} \quad \hat{\mathbf{X}}_{ar}^{(*)} = \left(\mathbf{I} - \mathbf{K}_{ar}^{(*)}\mathbf{H}_r^{(*)}(\hat{\mathbf{x}}_a^{(*)})\right)\hat{\mathbf{X}}_a^{(*)} \tag{273}$$

$$\mathbf{K}_{ar}^{(*)} = \hat{\mathbf{X}}_{ar}^{(*)}\mathbf{H}_r^{(*)T}(\hat{\mathbf{x}}_a^{(*)})\left(\mathbf{H}_r^{(*)}(\hat{\mathbf{x}}_a^{(*)})\hat{\mathbf{X}}_{ar}^{(*)}\mathbf{H}_r^{(*)T}(\hat{\mathbf{x}}_a^{(*)}) + \sigma_R^2\right)^{-1}. \tag{274}$$

Note that $\hat{r}^-$ appears in the first expression of (273) because $\hat{\xi}_a^{(*)} = (r^-/r_a)\hat{\xi}_a$ implies that $\hat{r}_a^{(*)} = \hat{r}^-$. Figure 6-13 illustrates the estimates of this case, along with those of the POT shown above. Figure 6-14 provides the position errors.

6.4 Comparison of the 2DOV CV EKF, POT Cases, and basic EPOT

Figure 6-17 and Figure 6-18 provide the sample mean-errors and sample standard deviations of the M-PLKF, EKF, and POT position estimates that were shown above (but now 500 Monte Carlo trials are being used). And the sample mean-errors and sample standard deviations of the EKF, POT, and EPOT cases are shown in Figure 6-19 and Figure 6-20.

Note that in Figure 6-17 and Figure 6-18 the EKF estimates are generally the worst, while the EKF with the POT is generally the best. (In the DCCM/ UCCM literature, the EKF usually provides the worst performance – there the EKF with the POT is ignored.) However, in Figure 6-19 and Figure 6-20 the EPOT is the worst. This performance issue of the EPOT shall be analyzed in the next Chapter; and there a remedy will be provided – accordingly, the version used here is dubbed the *Basic*-EPOT (B-EPOT).

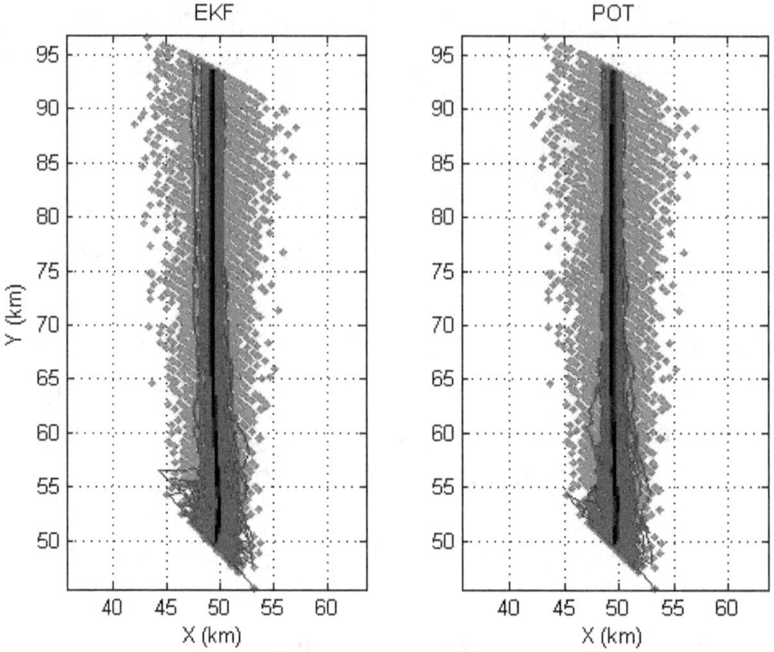

Figure 6-13 Comparison of EKF and POT (in Euclidean space)

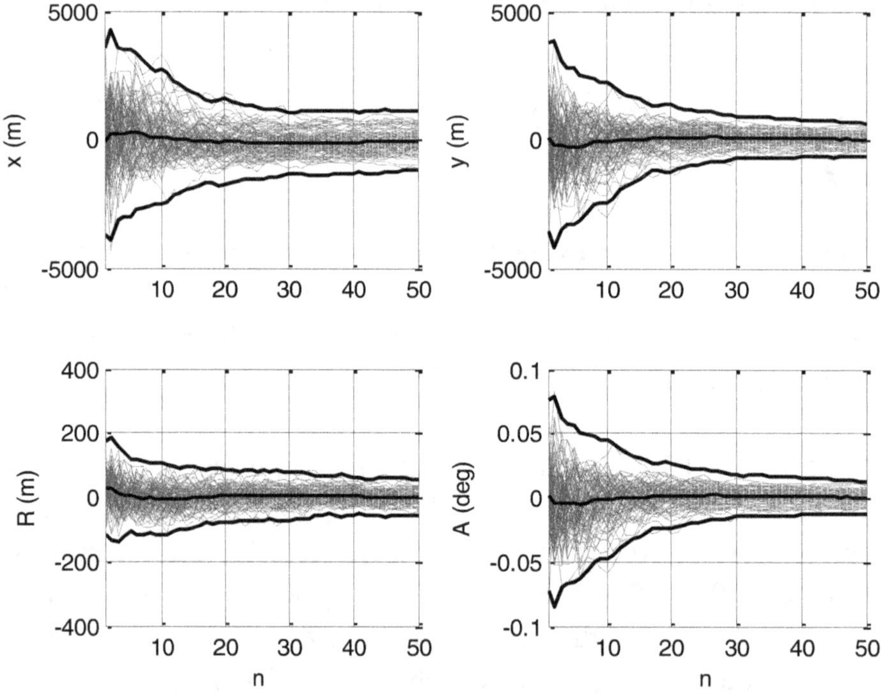

Figure 6-14 POT position estimation errors (2DOF CV case)

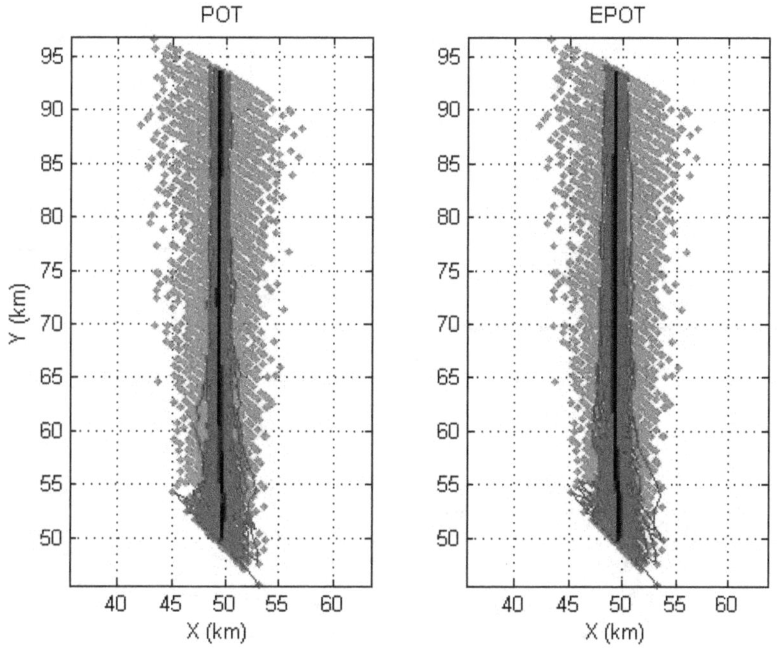

Figure 6-15 Comparison of POT and Basic EPOT (in Euclidean space)

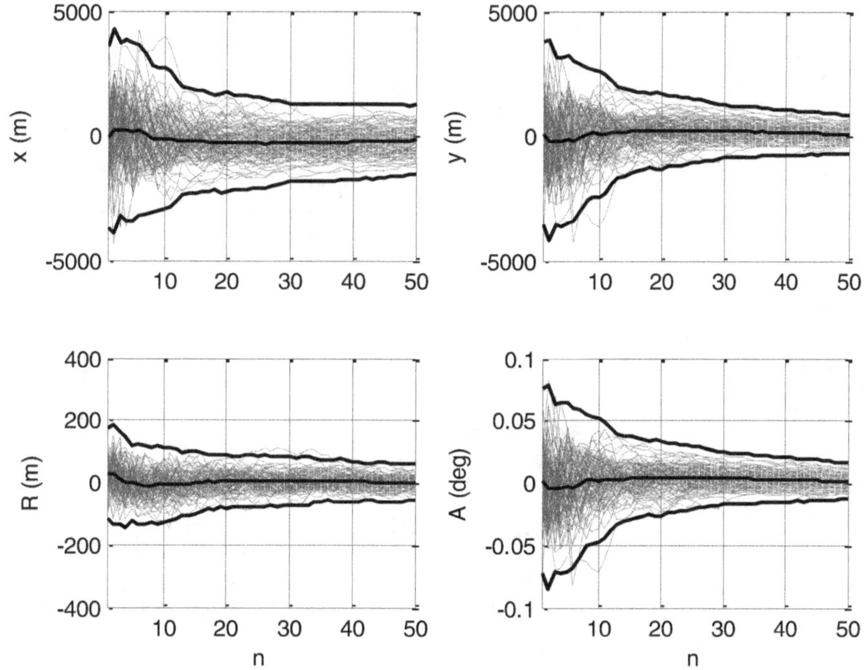

Figure 6-16 B-EPOT position estimation errors (2DOF CV case)

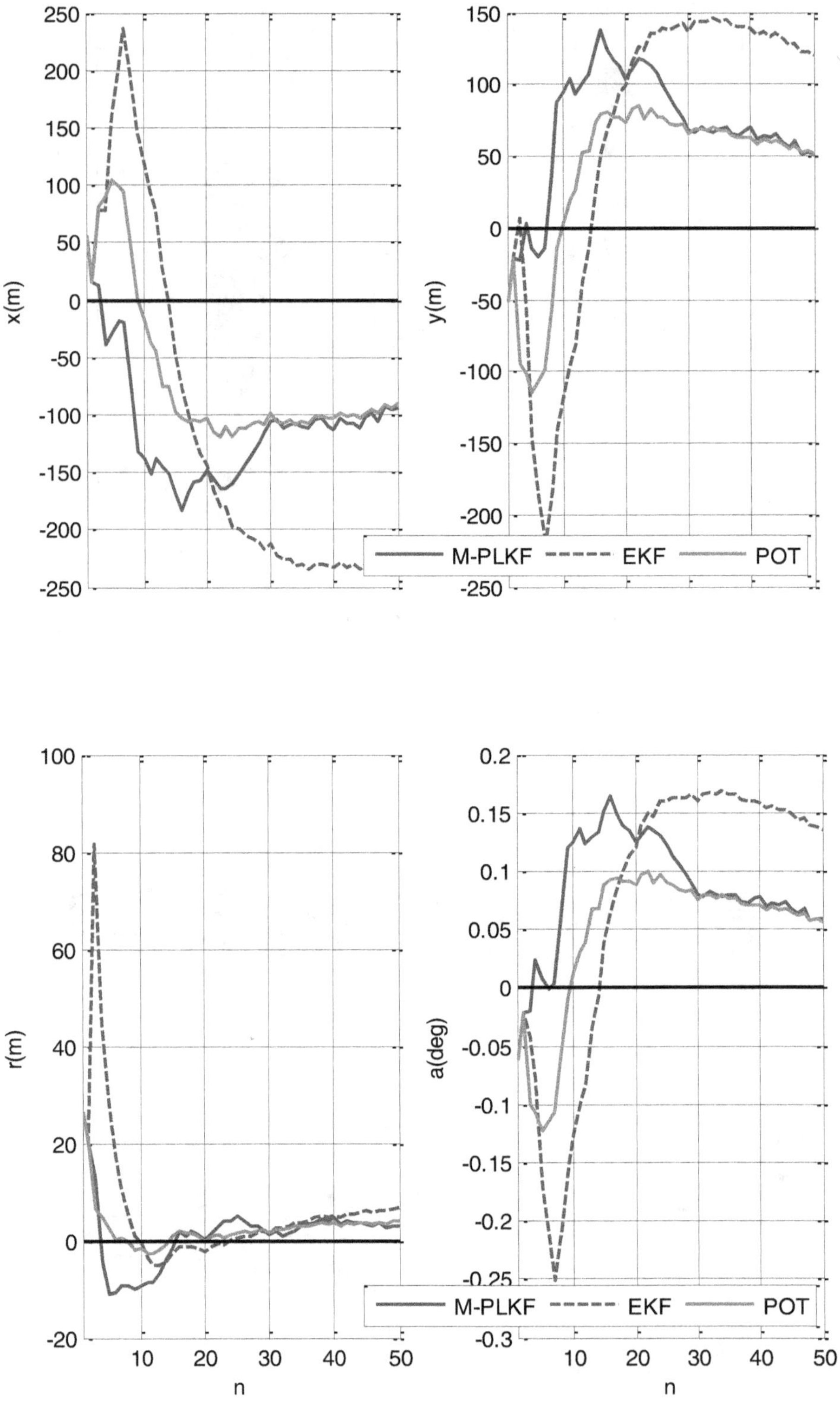

Figure 6-17 M-PLKF, EKF, and POT errors sample means (CV 2DOF)

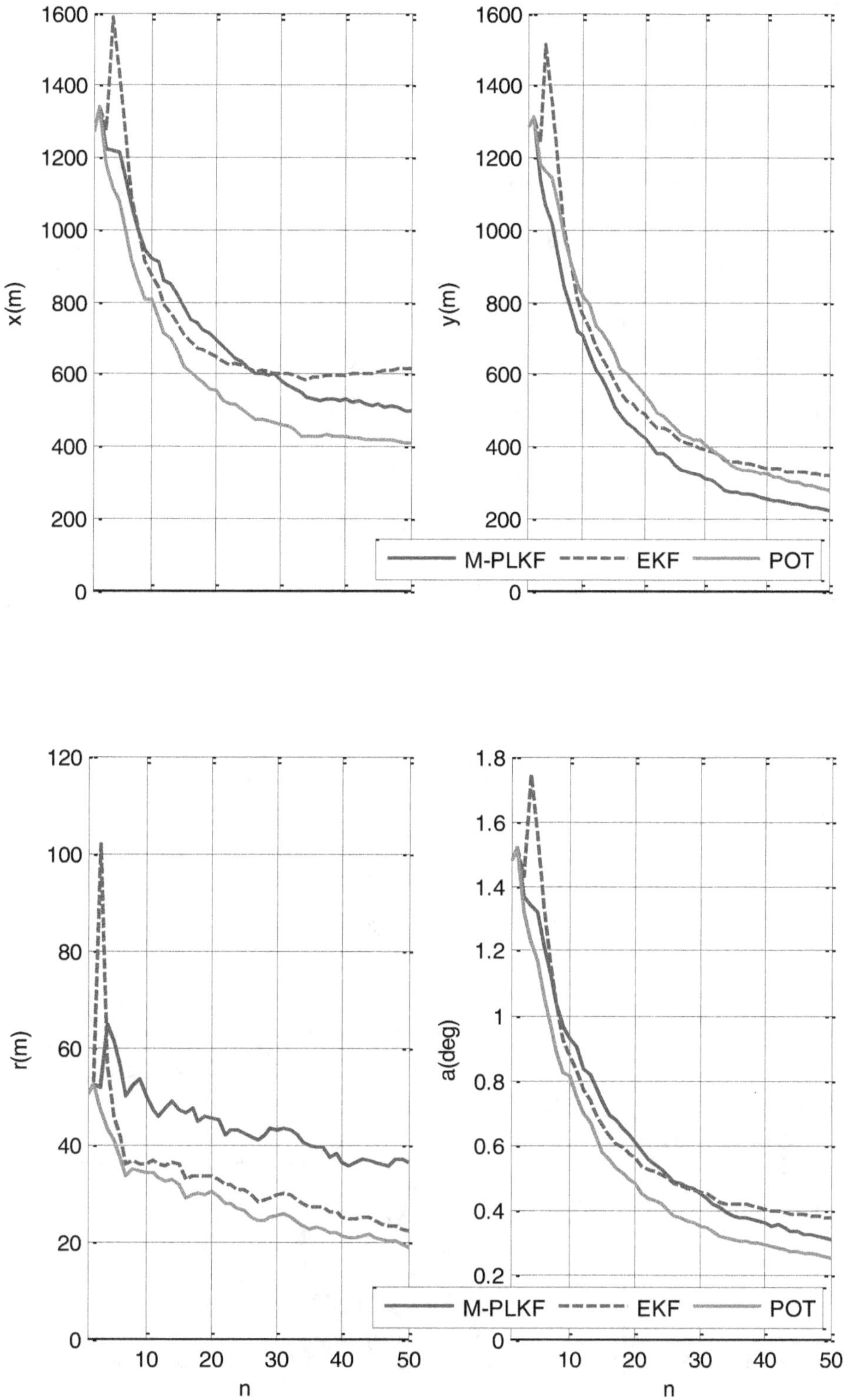

Figure 6-18 M-PLKF, EKF, and POT sample standard deviations (CV 2DOF)

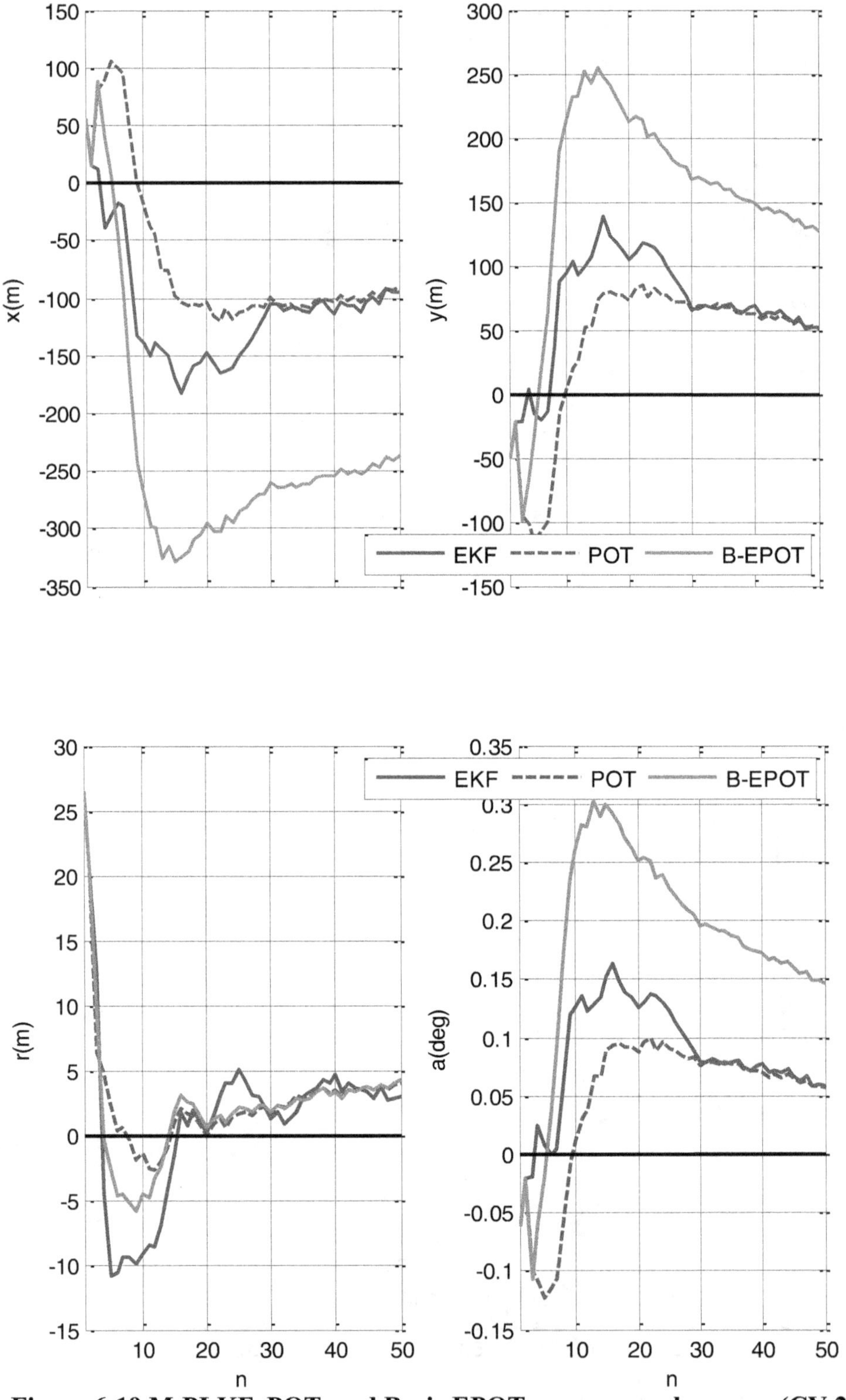

Figure 6-19 M-PLKF, POT, and Basic EPOT errors sample means (CV 2DOF)

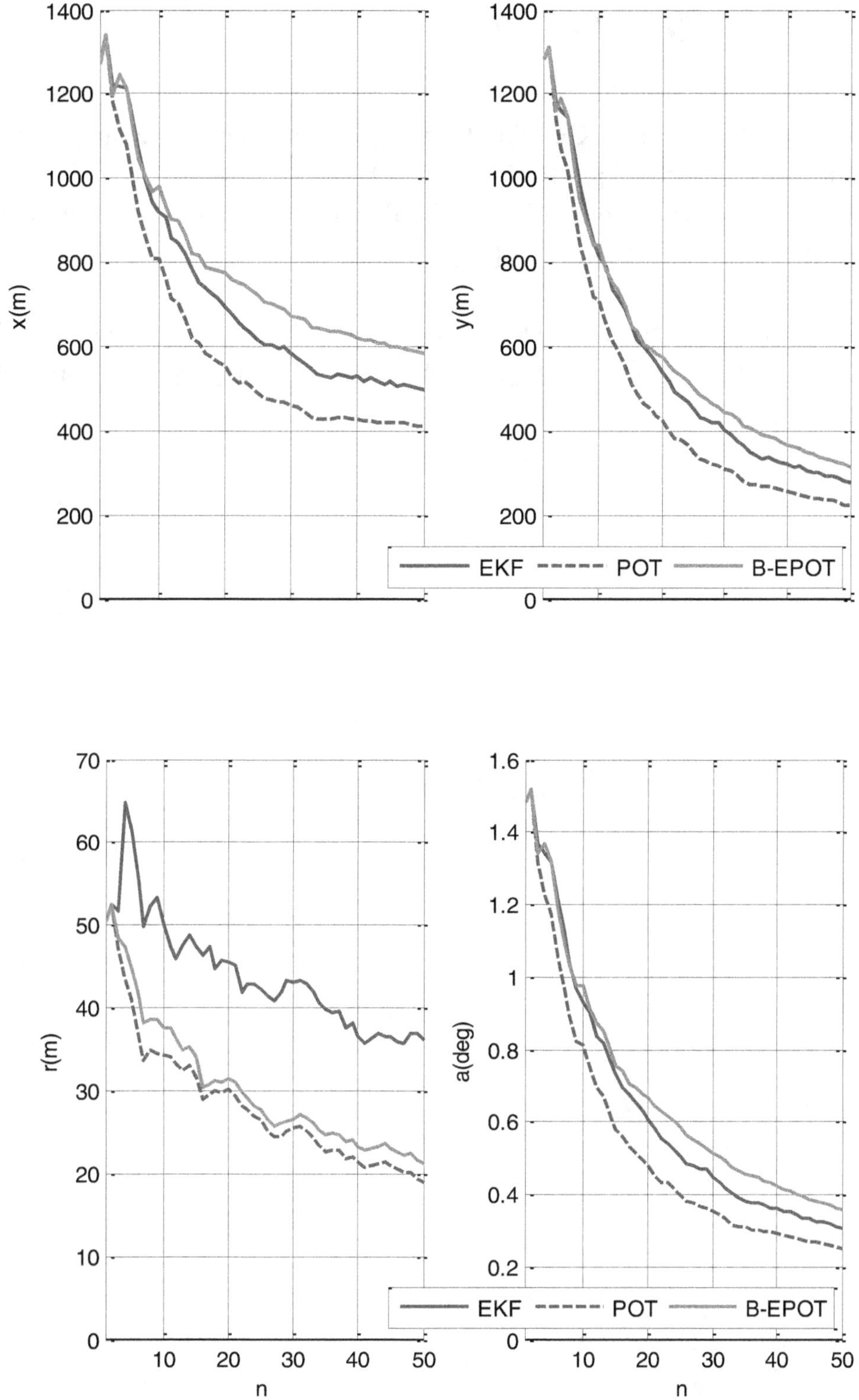

Figure 6-20 M-PLKF, POT, and Basic EPOT sample standard deviations (CV 2DOF)

6.5 Chapter 6 References

[1] D. Lerro and T. Bar-Shalom, "Tracking with Debiased Consistent Converted Measurements versus EKF," in Transactions on Aerospace and Electronic Systems, IEEE AES-29, pp. 1015-1022 (1993).

[2] W. Mei and Y. Bar-Shalom, "Unbiased Kalman Filter Using Converted Measurements: Revisit," in Signal and Data Processing of Small Targets, SPIE Vol. 7445 (2009).

[3] F. Daum and R. Fitzgerald, "Decoupled Kalman filters for Phased Array Radar Tracking," in IEEE Transactions on Automatic Control, Vol. 28 pp. 269-283 (1983).

7

The Position-Velocity Consistency Constraint

In Chapter 5, the CP case, the EPOT was seen to perform better than the POT, but in Chapter 6, where the object was moving the POT appeared to be better. Here it is shown in the latter case the EPOT's updates of position and velocity were inconsistent. Accordingly, a remedy for that inconsistency shall be provided. And it will then be shown that EPOT then overcomes the fundamental limitation POT: the measurement components of a detection may be used recursively in any order to update a track with virtually the same results, using the EKF. [1]

As in Chapter 5, the scalar-weight case will be analyzed first, followed by the matrix-weight case. And, as in Chapter 6, the state vector is $\mathbf{x}^T = (\boldsymbol{\xi}^T, \dot{\boldsymbol{\xi}}^T)$ with $\boldsymbol{\xi}^T = (x, y)$ and $\dot{\boldsymbol{\xi}}^T = (\dot{x}, \dot{y})$. The coordinate transformations are $\boldsymbol{\rho} = \mathbf{h}(\boldsymbol{\xi})$ and $\boldsymbol{\xi} = \mathbf{h}^{-1}(\boldsymbol{\rho})$ with $\boldsymbol{\rho}^T = (r, a)$ and $\mathbf{H}(\mathbf{x}) = \mathbf{J}(\boldsymbol{\xi}) \otimes (1,\ 0)$. For convenience, the EKF update equations are summarized as follows. Given $(\hat{\mathbf{x}}^-; \hat{\mathbf{X}}^-)$ and $(\overline{\boldsymbol{\rho}}; \boldsymbol{\Sigma}_R)$,

$$\hat{\mathbf{x}} = \hat{\mathbf{x}}^- + \mathbf{K}\left(\overline{\boldsymbol{\rho}} - \hat{\boldsymbol{\rho}}^-\right) \text{ and } \hat{\mathbf{X}} = \left(\mathbf{I} - \mathbf{K}\mathbf{H}(\hat{\mathbf{x}}^-)\right)\hat{\mathbf{X}}^- \tag{275}$$

with

$$\mathbf{K} = \hat{\mathbf{X}}^{-}\mathbf{H}^{T}(\hat{\mathbf{x}}^{-})\left(\mathbf{H}(\hat{\mathbf{x}}^{-})\hat{\mathbf{X}}^{-}\mathbf{H}^{T}(\hat{\mathbf{x}}^{-}) + \boldsymbol{\Sigma}_{R}\right)^{-1}. \tag{276}$$

And the alternate form for updating the estimate is

$$\hat{\mathbf{x}} = \hat{\mathbf{x}}^{-} + \hat{\mathbf{X}}\mathbf{H}^{T}(\hat{\mathbf{x}}^{-})\boldsymbol{\Sigma}_{R}^{-1}\left(\bar{\boldsymbol{\rho}} - \hat{\boldsymbol{\rho}}^{-}\right). \tag{277}$$

7.1 The Scalar-Weight Consistent EPOT

Recall that the scalar-weight EKF update is

$$\hat{\mathbf{x}} = \hat{\mathbf{x}}^{-} + \frac{\bar{w}}{\hat{w}}\mathbf{J}^{-1}(\hat{\mathbf{r}}^{-})\left(\bar{\boldsymbol{\rho}} - \hat{\boldsymbol{\rho}}^{-}\right) \text{ and } \hat{w} = \hat{w}^{-} + \bar{w}. \tag{278}$$

Here, to emphasize its geometrical aspects, that position update is written

$$\hat{\xi} = \hat{\xi}^{-} + r_{\Delta}\mathbf{e}_{\parallel}(\hat{a}^{-}) + \hat{r}^{-}a_{\Delta}\mathbf{e}_{\perp}(\hat{a}^{-}) \tag{279}$$

with $\mathbf{e}_{\parallel} \equiv \mathbf{e}_{r}$ and $\mathbf{e}_{\perp} \equiv \mathbf{e}_{a}$,

$$\mathbf{e}_{\parallel}^{T}(a) \equiv (\cos a,\ \sin a) \text{ and } \mathbf{e}_{\perp}^{T}(a) \equiv (-\sin a,\ \cos a), \tag{280}$$

and

$$r_{\Delta} = \frac{\bar{w}}{\hat{w}}(\bar{r} - \hat{r}^{-}) \text{ and } a_{\Delta} = \frac{\bar{w}}{\hat{w}}(\bar{a} - \hat{a}^{-}). \tag{281}$$

In the CV case, the scalar-weight EKF update is

$$\begin{bmatrix} \hat{\xi}_{CV}^{(s)} \\ \hat{\dot{\xi}}_{CV}^{(s)} \end{bmatrix} = \begin{bmatrix} \hat{\xi}^{-} \\ \hat{\dot{\xi}}^{-} \end{bmatrix} + \mathbf{K}_{CV}^{(s)}\begin{bmatrix} \bar{r} - \hat{r}^{-} \\ \bar{a} - \hat{a}^{-} \end{bmatrix}, \tag{282}$$

with

$$\mathbf{K}_{CV}^{(s)} = \mathbf{J}^{-1}(\hat{\boldsymbol{\rho}}^{-}) \otimes \begin{bmatrix} \alpha \\ \beta/\tau \end{bmatrix}, \tag{283}$$

where $\mathbf{J}^{-1}(\boldsymbol{\rho}) = \left[\mathbf{e}_{\parallel}(a) \quad r\mathbf{e}_{\perp}(a)\right]$. The details for determining α and β were given in

Chapter 6. There, using $m = n+2$, the gains were shown to be

$$\alpha_n = 2(2m-1)/m(m+1) \text{ and } \beta_n = 6/m(m+1)\tau . \tag{284}$$

In (283) these gains are being used, but without the indices to simplify the notation. For the remainder of this section the superscript "(s)" and subscript "CV" shall be dropped.

Now write the scalar-weight EKF update determined by (282) and (283) as

$$\hat{\xi} = \hat{\xi}^- + \alpha \begin{bmatrix} \cos\hat{a}^- & -\hat{r}^- \sin\hat{a}^- \\ \sin\hat{a}^- & +\hat{r}^- \cos\hat{a}^- \end{bmatrix} \begin{bmatrix} \bar{r} - \hat{r}^- \\ \bar{a} - \hat{a}^- \end{bmatrix} \tag{285}$$

and

$$\hat{\dot{\xi}} = \hat{\dot{\xi}}^- + (\beta/\tau) \begin{bmatrix} \cos\hat{a}^- & -\hat{r}^- \sin\hat{a}^- \\ \sin\hat{a}^- & +\hat{r}^- \cos\hat{a}^- \end{bmatrix} \begin{bmatrix} \bar{r} - \hat{r}^- \\ \bar{a} - \hat{a}^- \end{bmatrix}. \tag{286}$$

And let

$$\Delta r^- \equiv \alpha(\bar{r} - \hat{r}^-) \text{ and } \Delta a^- \equiv \alpha(\bar{a} - \hat{a}^-) \tag{287}$$

and

$$\Delta \dot{r}^- \equiv (\beta/\tau)(\bar{r} - \hat{r}^-) \text{ and } \Delta \dot{a}^- \equiv (\beta/\tau)(\bar{a} - \hat{a}^-). \tag{288}$$

Finally, write (285) and (286) together as

$$\begin{bmatrix} \hat{\xi} \\ \hat{\dot{\xi}} \end{bmatrix} = \begin{bmatrix} \hat{\xi}^- \\ \hat{\dot{\xi}}^- \end{bmatrix} + \begin{bmatrix} \Delta r^- & \hat{r}^- \Delta a^- \\ \Delta \dot{r}^- & \hat{r}^- \Delta \dot{a}^- \end{bmatrix} \begin{bmatrix} \mathbf{e}_{\parallel}(\hat{a}^-) \\ \mathbf{e}_{\perp}(\hat{a}^-) \end{bmatrix}. \tag{289}$$

Thus, using $\mathbf{x}^T = (\xi^T, \dot{\xi}^T)$,

$$\hat{\mathbf{x}} = \hat{\mathbf{x}}^- + \Delta r^- \mathbf{e}_{\parallel}(\hat{a}^-) + \hat{r}^- \Delta \dot{a}^- \mathbf{e}_{\perp}(\hat{a}^-). \tag{290}$$

Next, recall the affine space $\mathbf{A}$. Given $\xi \in \boldsymbol{X}$ and $\rho \in \boldsymbol{R}$, with $\rho = \mathbf{h}(\xi)$, the column vectors ξ and ρ represent the same point in $\mathbb{E}$. Their corresponding (abstract) vectors in $\mathbf{A}$ are the same, $\xi = \rho$. That is, using equation (5) of Chapter 2 with $r = \sqrt{x^2 + y^2}$ and $a = \arctan(y, x)$,

$$\xi = x\boldsymbol{e}_x + x\boldsymbol{e}_y = r\boldsymbol{e}_{\parallel}(a) + 0\boldsymbol{e}_{\perp}(a) = \rho\,. \tag{291}$$

Thus, in **A** the position and velocity updates given by (285) and (286) are simply

$$\hat{\xi} = \hat{\xi}^- + \Delta r^- \boldsymbol{e}_{\parallel}^- + \hat{r}^- \Delta a^- \boldsymbol{e}_{\perp}^- \;\text{ and }\; \hat{\dot{\xi}} = \hat{\dot{\xi}}^- + \Delta\dot{r}^- \boldsymbol{e}_{\parallel}^- + \hat{r}^- \Delta\dot{a}^- \boldsymbol{e}_{\perp}^-\,, \tag{292}$$

where, for convenience, $\boldsymbol{e}_{\parallel}^- \equiv \boldsymbol{e}_{\parallel}(\hat{a}^-)$ and $\boldsymbol{e}_{\perp}^- \equiv \boldsymbol{e}_{\perp}(\hat{a}^-)$. Finally, in **A** the update equation corresponding to (282) and (289) is

$$\begin{bmatrix} \hat{\xi} \\ \hat{\dot{\xi}} \end{bmatrix} = \begin{bmatrix} \hat{\xi}^- \\ \hat{\dot{\xi}}^- \end{bmatrix} + \begin{bmatrix} \Delta r^- & \hat{r}^- \Delta a^- \\ \Delta\dot{r}^- & \hat{r}^- \Delta\dot{a}^- \end{bmatrix} \begin{bmatrix} \boldsymbol{e}_{\parallel}^- \\ \boldsymbol{e}_{\perp}^- \end{bmatrix}. \tag{293}$$

Figure 7-1 depicts the geometrical aspect of this EKF update of position in **A**, along with the corresponding LKF update (the velocity update will be discussed shortly). The azimuth-first-range-last *sequential* EKF update is

$$\hat{\xi}_a = \hat{\xi}^- + \hat{r}^- \Delta a^- \boldsymbol{e}_{\perp}^- \;\text{ and }\; \hat{\xi}_{\text{EKF}} = \hat{\xi}_a + \Delta r^- \boldsymbol{e}_{\parallel}^-\,. \tag{294}$$

The sequential range-first-azimuth-last update provides the same result, $\hat{\xi}_r = \hat{\xi}^- + \hat{r}^- \Delta r^- \boldsymbol{e}_{\parallel}^-$ and $\hat{\xi}_{\text{EKF}} = \hat{\xi}_r + \hat{r}^- \Delta a^- \boldsymbol{e}_{\perp}^-$. Note that the LKF position update is simply the composition of rotating $\hat{\xi}^-$ by the angle Δa^-, and the translation along the resulting radial by Δr^- (and these two operations of rotation and translation commute).

Figure 7-2 illustrates the corresponding updates determined by the POT and EPOT. For the POT, the azimuth-first update is the same as in (294); but its recursive range-last update employs $\hat{r}_a = \sqrt{\hat{x}_a^2 + \hat{y}_a^2}$ and $\hat{a}_a = \arctan(\hat{y}_a, \hat{x}_a)$. That is,

$$\hat{\xi}_{\text{POT}} = \hat{\xi}_a + \Delta r_a \boldsymbol{e}_{\parallel}^{(a)}\,, \tag{295}$$

where

$$\Delta r_a \equiv \alpha(\overline{r} - \hat{r}_a) \;\text{ and }\; \boldsymbol{e}_{\parallel}^{(a)} \equiv \boldsymbol{e}_{\parallel}(\hat{a}_a)\,. \tag{296}$$

Similarly, given $\hat{\xi}_a$, the range-last position update determined by the EPOT is

$$\hat{\xi}_{\text{EPOT}} = \hat{\xi}_a^{(*)} + \Delta r^- \boldsymbol{e}_{\parallel}^{(a)}\,, \tag{297}$$

where $\hat{\xi}_a^{(*)} \equiv (\hat{r}^- / \hat{r}_a)\hat{\xi}_a$, with Δr^- determined by (287).

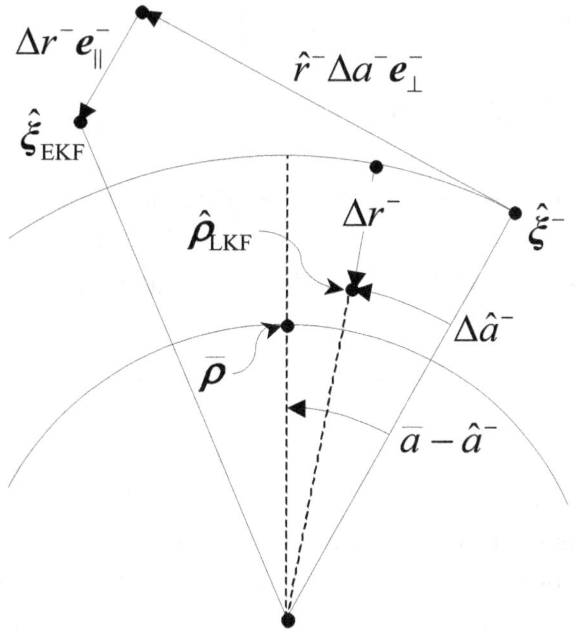

Figure 7-1 Geometry of the EKF position update

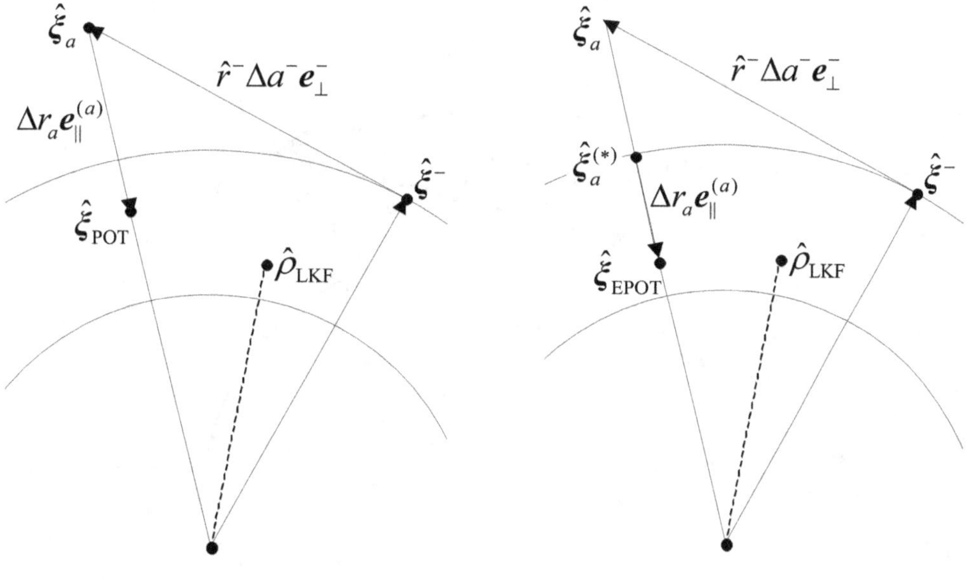

Figure 7-2 Geometry of the POT and EPOT position updates

Note that (285) and (286) are related by $\bar{\boldsymbol{\rho}} - \hat{\boldsymbol{\rho}}^-$, namely,

$$\hat{\xi} = \hat{\xi}^- + \alpha \mathbf{J}^{-1}(\hat{\boldsymbol{\rho}}^-)\left(\bar{\boldsymbol{\rho}} - \hat{\boldsymbol{\rho}}^-\right) \quad \text{and} \quad \hat{\dot{\xi}} = \hat{\dot{\xi}}^- + (\beta/\tau)\mathbf{J}^{-1}(\hat{\boldsymbol{\rho}}^-)\left(\bar{\boldsymbol{\rho}} - \hat{\boldsymbol{\rho}}^-\right). \tag{298}$$

In particular, given the residual $\bar{\boldsymbol{\rho}} - \hat{\boldsymbol{\rho}}^-$, the update determines a position differential, $\hat{\xi} - \hat{\xi}^-$, and a velocity differential, $\hat{\dot{\xi}} - \hat{\dot{\xi}}^-$. Figure 7-3 shows the respective position differentials for the LKF, EKF, POT, and EPOT updates that were illustrated above in Figure 7-1 and Figure 7-2.

Next, solve for $\hat{\xi} - \hat{\xi}^-$ in the first expression in (298),

$$\bar{\boldsymbol{\rho}} - \hat{\boldsymbol{\rho}}^- = (1/\alpha)\mathbf{J}(\hat{\xi}^-)\left(\hat{\xi} - \hat{\xi}^-\right). \tag{299}$$

And then substitute this result into the second expression in (298). Also, recall that $\mathbf{I} = \mathbf{J}(\hat{\boldsymbol{\rho}}^-)\mathbf{J}^{-1}(\hat{\boldsymbol{\rho}}^-)$ is an identity when $\hat{\boldsymbol{\rho}}^- \neq \mathbf{0}$. Thus, the velocity differential determined by the second expression in (298) is

$$\hat{\dot{\xi}} - \hat{\dot{\xi}}^- = (\beta/\tau\alpha)\left(\hat{\xi} - \hat{\xi}^-\right). \tag{300}$$

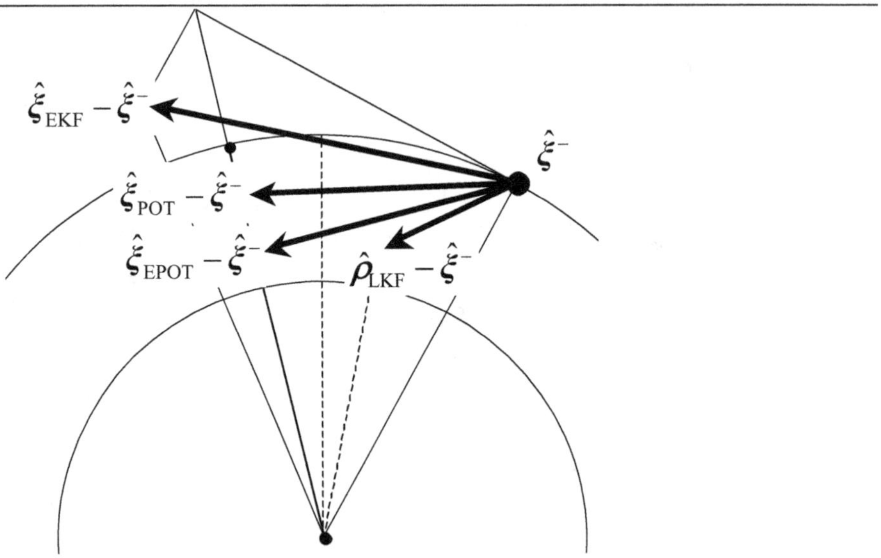

Figure 7-3 Position differentials of the updates

This last expression defines the *position-velocity consistency condition* for the update. Such is simply a linear relation between differentials, $\hat{\xi}-\hat{\xi}^-$ and $\hat{\dot{\xi}}-\hat{\dot{\xi}}^-$. Indeed, in the scalar-weight case being used in this section, $\hat{\xi}-\hat{\xi}^-$ and $\hat{\dot{\xi}}-\hat{\dot{\xi}}^-$ are related by the scalar $(\beta/\tau\alpha)$, which is independent of both $\bar{\mathbf{p}}$ and $\hat{\xi}^-$. The LKF, EKF, and POT cases inherently satisfy the position-velocity consistency condition defined above, but the "Basic-EPOT" does not (the Basic-EPOT modifies $\hat{\xi}-\hat{\xi}^-$, without changing $\hat{\dot{\xi}}-\hat{\dot{\xi}}^-$).

Fortunately, the EPOT can easily be made to satisfy (300) by first updating the position estimate (using the Basic-EPOT), and then updating the velocity using

$$\hat{\dot{\xi}}_{\text{EPOT}} \equiv \hat{\dot{\xi}}^- + (\beta/\alpha\tau)\left(\hat{\xi}_{\text{EPOT}} - \hat{\xi}^-\right). \quad (301)$$

Figure 7-4 and Figure 7-5 illustrate the performance of the EPOT with this *velocity-consistent* scalar-weight update, along with the "debiased" (UCCM) scalar-weight PLKF and the scalar-weight POT. Figure 7-4 provides the sample means of the estimation errors and Figure 7-5 provides the sample standard deviations of those errors. Here the same scenario used in Chapter 6 is being reused, but with $\sigma_A = 2.5$ instead of $\sigma_A = 1.5$. Note that the respective sample means are more or less the same, except for range – those of the "debiased" S-PLKF are worse than those of the S-POT and S-EPOT. Note also that the sample standard deviations of the S-PLKF, S-POT and S-EPOT position estimates are now indistinguishable, except for range: the S-PLKF range estimates are the worst while the S-EPOT range estimates are now the best.

7.2 The Matrix-Weight Consistent EPOT Case

In this Section the *position-velocity consistency condition* for the matrix -weight EPOT update is provided, and then illustrated. First, given the CV form of the updated estimate $\hat{\mathbf{x}}^T = (\hat{\xi}^T, \hat{\dot{\xi}}^T)$, write its associated covariance matrix as

$$\mathbf{X} = \begin{bmatrix} \mathbf{X}_{\xi\xi} & \mathbf{X}_{\xi\dot{\xi}} \\ \mathbf{X}_{\dot{\xi}\xi} & \mathbf{X}_{\dot{\xi}\dot{\xi}} \end{bmatrix}.$$

Here each subscripted "$\mathbf{X}$" is a 2×2 sub-matrix; and $\mathbf{X}_{\xi\xi}$ is positive definite since $\hat{\mathbf{X}}$ is positive definite by assumption. To analyze the matrix-weight case, an approach similar to that taken in the previous section will be followed.

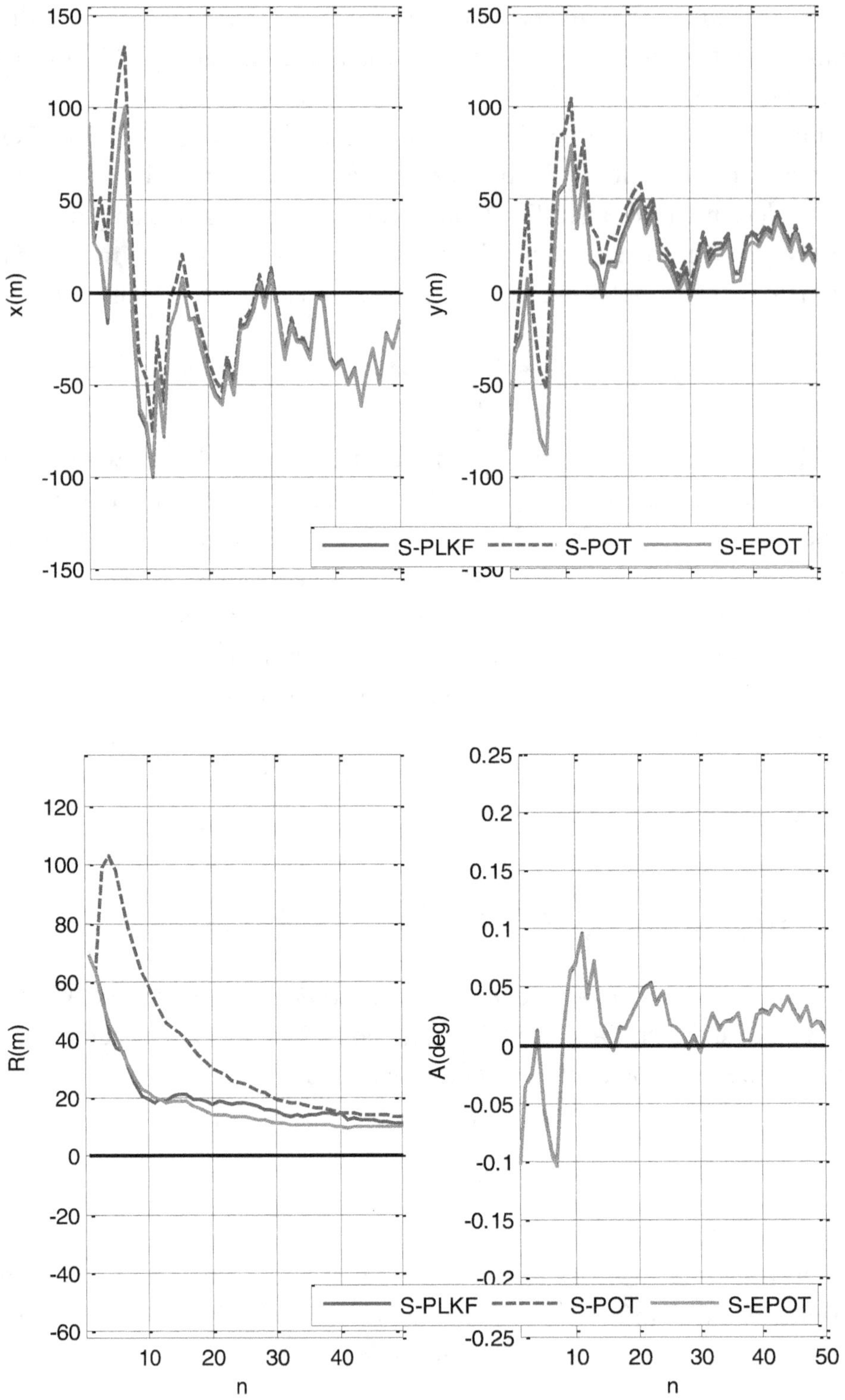

Figure 7-4 PLKF, POT, and EPOT sample means (scalar-weight case)

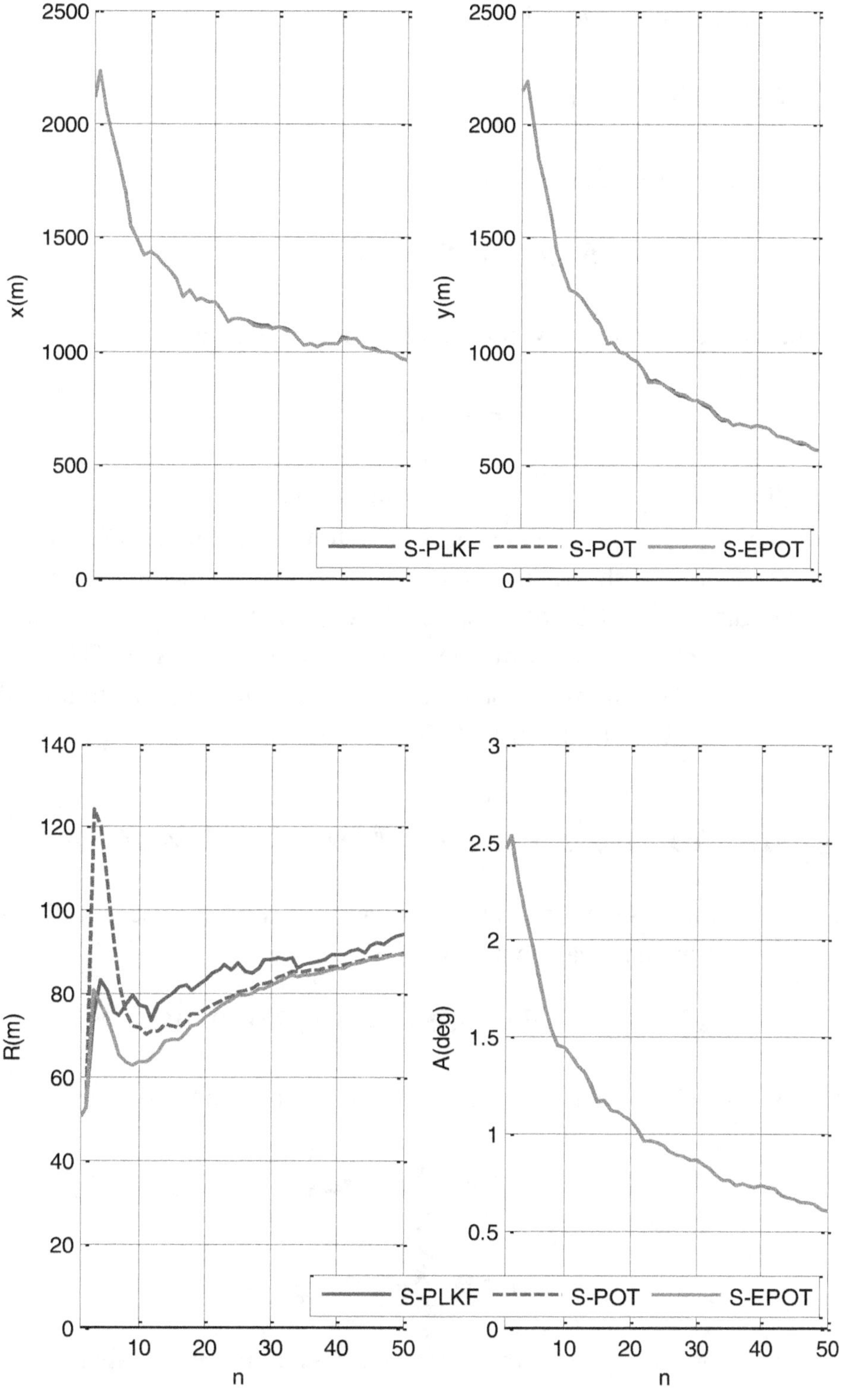

Figure 7-5 PLKF, POT, and EPOT sample standard deviations (scalar-weight case)

Now write the position and velocity updates respectively as

$$\hat{\xi}-\hat{\xi}^{-}=\mathbf{X}_{\xi\xi}\mathbf{J}^{T}\boldsymbol{\Sigma}_{R}^{-1}(\bar{\boldsymbol{\rho}}-\hat{\boldsymbol{\rho}}^{-}) \text{ and } \hat{\dot{\xi}}-\hat{\dot{\xi}}^{-}=\mathbf{X}_{\dot{\xi}\xi}\mathbf{J}^{T}\boldsymbol{\Sigma}_{R}^{-1}(\bar{\boldsymbol{\rho}}-\hat{\boldsymbol{\rho}}^{-})\,. \tag{302}$$

As before, the first expression in (302) provides

$$\bar{\boldsymbol{\rho}}-\hat{\boldsymbol{\rho}}^{-}=\boldsymbol{\Sigma}_{R}\mathbf{J}^{-T}\mathbf{X}_{\xi\xi}^{-1}\left(\hat{\xi}-\hat{\xi}^{-}\right). \tag{303}$$

And substituting this result into the second expression in (302) leads to

$$\hat{\dot{\xi}}-\hat{\dot{\xi}}^{-}=\mathbf{X}_{\dot{\xi}\xi}\mathbf{X}_{\xi\xi}^{-1}\left(\hat{\xi}-\hat{\xi}^{-}\right). \tag{304}$$

Such is the position-velocity consistency constraint for the matrix-weight case.

Figure 7-6 and Figure 7-7 provide the estimation errors of the *consistent* EPOT, defined in this section. Also shown are the corresponding errors M-PLKF and POT that were shown in Figures 26 and 27 of Chapter 6. Here, for the sake of comparison with those previous results, $\sigma_A = 1.5$. Now the mean-errors of the EPOT and POT are basically the same for range; and their sample standard deviations are mostly indistinguishable. The ones of the "debiased and consistent" M-PLKF are the worst.

7.3 Extension to Higher-Order Models of Motion

The (consistent) CV matrix-weight EPOT is extended to the CA case as follows. First, write the CA state vector and measurement model matrix as

$$\mathbf{x}^{T}=(\xi^{T},\ \dot{\xi}^{T},\ \ddot{\xi}^{T}) \text{ and } \mathbf{H}(\xi)=\mathbf{J}(\mathbf{x})\otimes(1,\ 0,\ 0)\,. \tag{305}$$

And write the associated covariance matrix of the updated estimate as

$$\mathbf{X}=\begin{bmatrix}\mathbf{X}_{\xi\xi} & \mathbf{X}_{\xi\dot{\xi}} & \mathbf{X}_{\xi\ddot{\xi}}\\ \mathbf{X}_{\dot{\xi}\xi} & \mathbf{X}_{\dot{\xi}\dot{\xi}} & \mathbf{X}_{\dot{\xi}\ddot{\xi}}\\ \mathbf{X}_{\ddot{\xi}\xi} & \mathbf{X}_{\ddot{\xi}\dot{\xi}} & \mathbf{X}_{\ddot{\xi}\ddot{\xi}}\end{bmatrix}.$$

The CA expression corresponding to the second expression in (302) is then

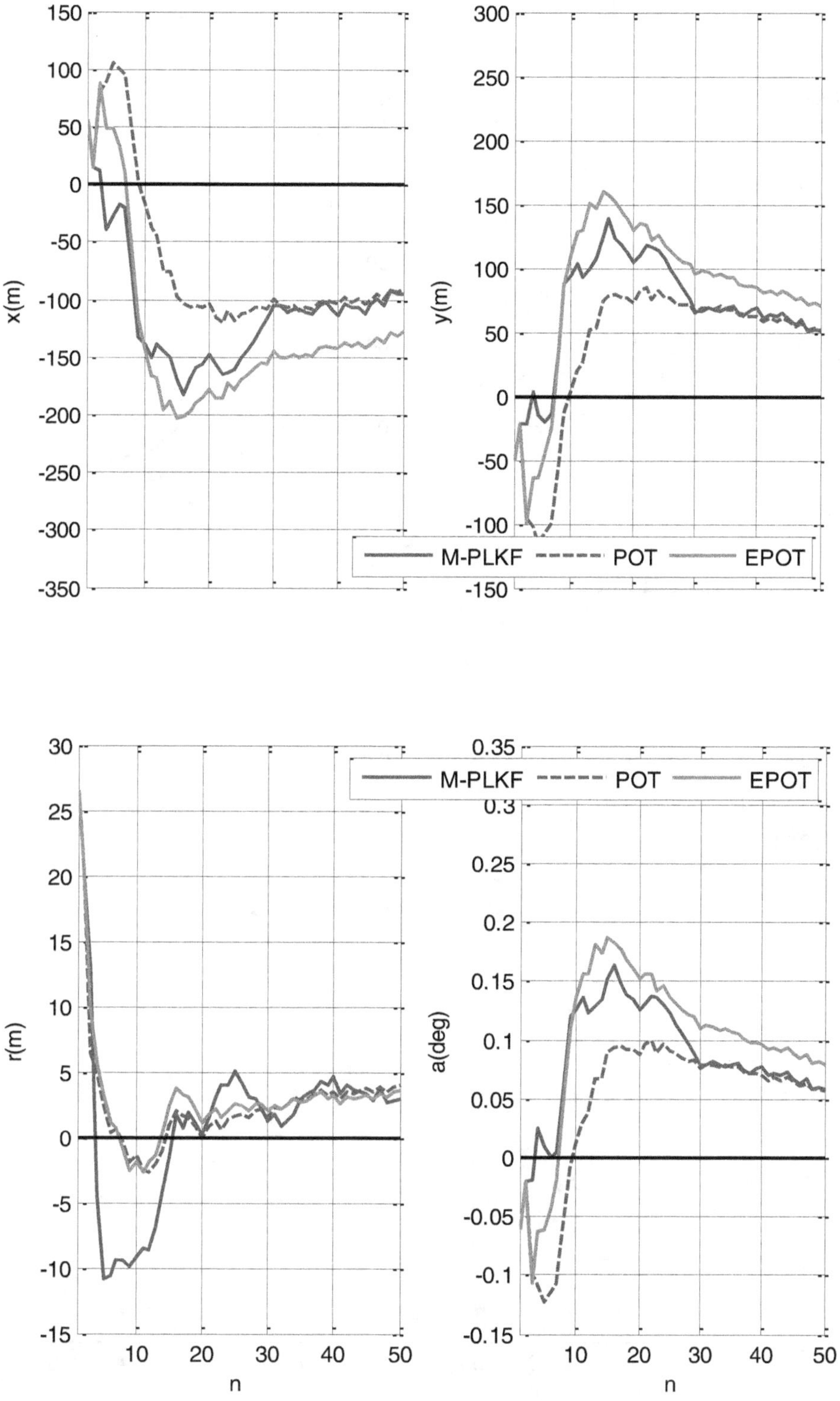

Figure 7-6 M-PLKF, POT, and "CV EPOT" sample mean errors (CV 2DOF)

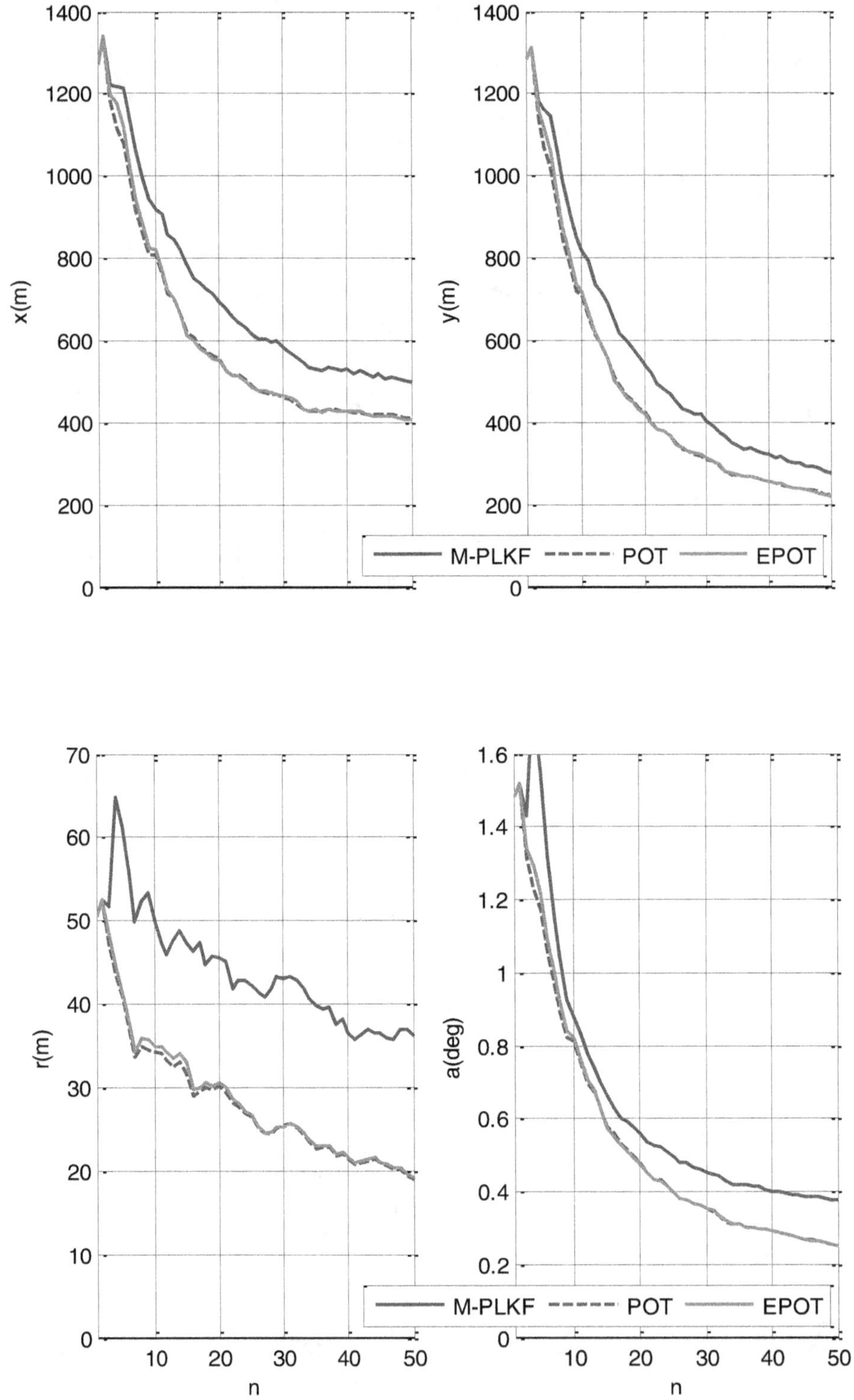

Figure 7-7 M-PLKF, POT, and "CV EPOT" sample standard deviations (CV 2DOF)

$$\begin{bmatrix} \hat{\dot{\xi}} - \hat{\dot{\xi}}^- \\ \hat{\ddot{\xi}} - \hat{\ddot{\xi}}^- \end{bmatrix} = \begin{bmatrix} \mathbf{X}_{\dot{\xi}\xi} \\ \mathbf{X}_{\ddot{\xi}\xi} \end{bmatrix} \mathbf{J}^T \boldsymbol{\Sigma}_R^{-1} (\overline{\boldsymbol{\rho}} - \hat{\boldsymbol{\rho}}^-). \tag{306}$$

And the expression for the CA case corresponding to (304) is

$$\begin{bmatrix} \hat{\dot{\xi}} - \hat{\dot{\xi}}^- \\ \hat{\ddot{\xi}} - \hat{\ddot{\xi}}^- \end{bmatrix} = \begin{bmatrix} \mathbf{X}_{\dot{\xi}\xi} \\ \mathbf{X}_{\ddot{\xi}\xi} \end{bmatrix} \mathbf{X}_{\xi\xi}^{-1} \left(\hat{\xi} - \hat{\xi}^- \right). \tag{307}$$

Finally, this *position-velocity-acceleration* consistency constraint readily generalizes to the pth-order case.

7.4 The EPOT's "Preferred Ordering"

The "consistent" EPOT abolishes the preferred ordering requirement in the POT. Figure 7-8 illustrates such using the sample mean-errors and "sample confidence intervals." The corresponding results for the Basic-EPOT used in Chapter 6, are provided by Figure 7-9.

7.5 Chapter 7 References

[1] D. M. Leskiw, "The Extended Preferred Ordering Theorem for Radar Tracking Using the Extended Kalman Filter", PhD Dissertation, Syracuse University (2012).

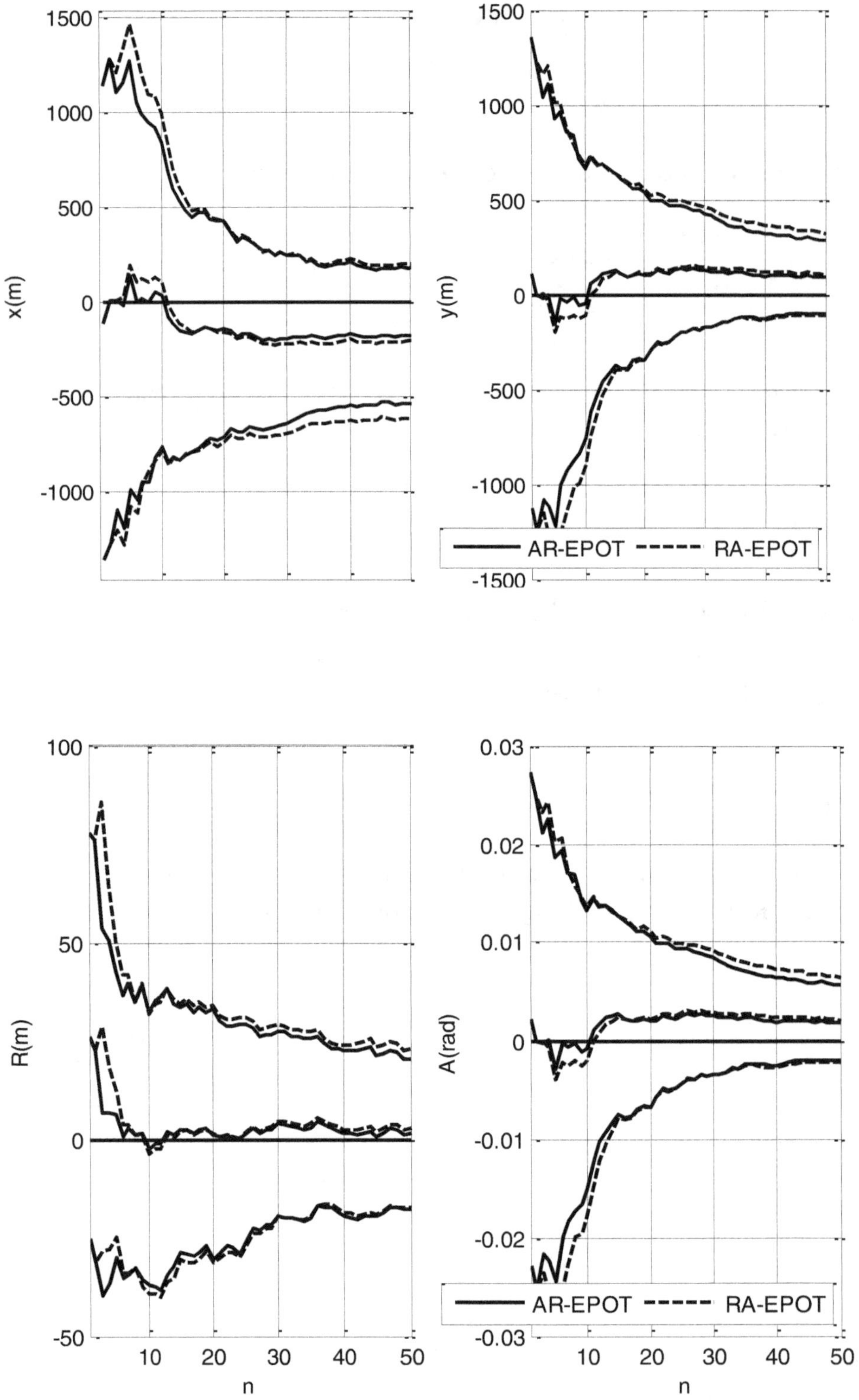

Figure 7-8 Comparison between the AR-EPOT and RA-EPOT (consistent case)

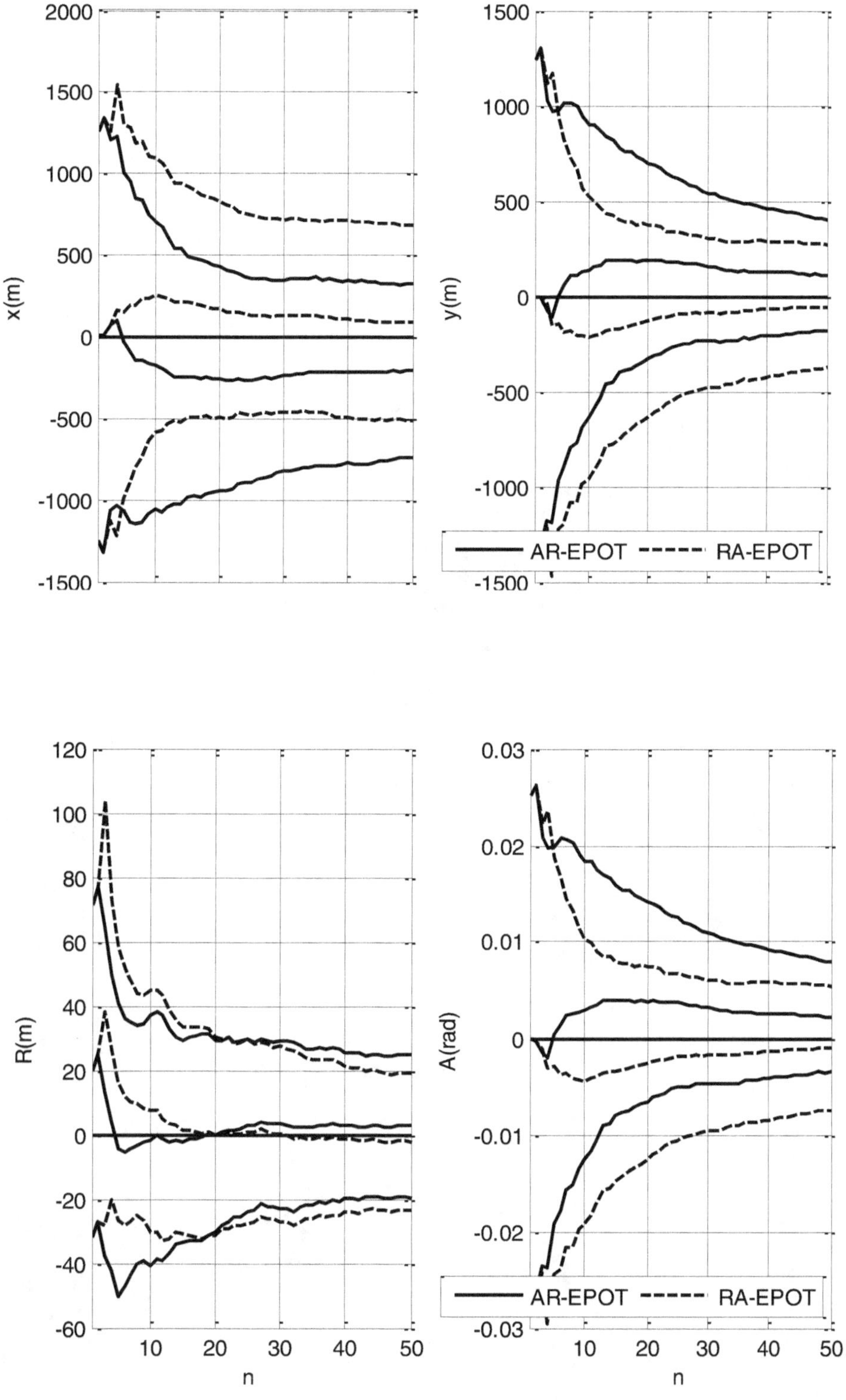

Figure 7-9 Comparison between the AR-EPOT and RA-EPOT (inconsistent case)

8

Concluding Remarks

8.1 Summary of this Research.

As mentioned in the introduction, most radar tracking practitioners prefer to use some variation of the PLKF, not the EKF. But the PLKF makes the estimates noisier, and its tracks are inherently biased. And we have shown that the leading "debiasing" methods for the PLKF can make the tracks even more biased. Fortunately, the POT eliminates most of the EKF errors: the EKF tracks under the POT are generally better than the corresponding PLKF ones – less biased and less noisy. The POT, however, has an obvious limitation: by definition, its tracks must be updated using azimuth first and range last. Also, the POT is not as effective at short ranges.

This work has provided a new version of the POT, dubbed the Extended-POT (EPOT). Not only is the EPOT more efficient than the POT at short ranges, it overcomes the fundamental limitation of the POT – the independent measurement components of a detection can now be used recursively by an EKF in any order, with virtually the same results.

The key results are summarized as follows.

- Definition of the Problem. In Chapter 3 a more stressing variation of the Julier and Uhlmann CP "exemplar" was used to show that PLKF tracks tend to be biased, and more noisy than those of the LKF. It was shown that two distinct PLKF bias cases exist, scalar- and matrix-weight, their biases having opposite sense to one another. The EKF performed better than the PLKF but was also seen to have a convergence problem.

- Analysis of the PLKF. In Chapter 4 it was shown that the UCCM "debiasing" can make the PLKF biases worse. And it was also shown that the popular approximations for the "consistent" covariance matrix can lead to "inconsistent" gain matrices. But a remedy for that was provided – and that modified UCCM was seen to be less biased, albeit slightly noisier.
- Analysis of the EKF. In Chapter 5, using the CP exemplar of Julier and Uhlmann it was shown that the POT is not efficient at very short ranges. A new method was introduced, to be dubbed the Basic-EPOT. Such was seen to perform much better than the POT, and better than the "debiased and consistent" PLKF, on the Julier and Uhlmann CP exemplar.
- Application to Radar Tracking. In Chapter 6 the CV exemplar of Lerro and Bar Shalom (from the DCCM/UCCM literature) was used to compare the effectiveness of the "debiased and consistent" PLKF with those of the EKF (with and without the POT and EPOT). There the EKF with the POT was best; the Basic-EPOT was worse than the POT.
- Analysis of the EPOT. In Chapter 7 the various methods that were illustrated in the previous Chapters were analyzed further and a certain position-velocity *consistency condition* was derived. The LKF, PLKF, EKF, and POT inherently satisfy that condition, while the CP version of the EPOT given in Chapter 5 did not. A remedy was then given for the CV case (and for the CA case); whereupon, the performance of the EPOT became comparable to that of the POT. Moreover, it was shown that that consistent EPOT overcomes the "preferred ordering" limitation of the POT.

8.2 Additional Remarks

8.2.1 The Effect of Non-Stationary Observation Geometries

Let the object in the UCCM CV exemplar be initially inbound, moving south (along an axis parallel to the *Y*-axis), such that its range at closest approach is one kilometer. Figure 8-1 illustrates this case for the EPOT (with $\sigma_A = 1.5$ and zero system noise), using 500 Monte Carlo trials. The left-hand side shows the detections and tracks in $\mathbb{E}$; and the right-hand side shows the detections and tracks converted back into radar coordinates. Note that the spread of the measurements shrinks in cross-range as the object approaches the radar and expands in cross-range as the object moves away. Figure 8-2 provides the corresponding *rms-error* (rmse) of the estimated position radar coordinates for the M-PLKF, POT, and EPOT, with the "debiased" PLKF using the true covariance matrices of the rectangular pseudo-measurements. (Here the *rmse* is defined as the square-root of the sum of the sample mean-error squared plus the sample variance, $\sqrt{\hat{\mu}^2 + \hat{\sigma}^2}$.) The corresponding results for $\sigma_A = 2.5$ degrees are provided in Figure 8-3.

Note that all the rmse's are monotone deceasing as more detections are processed (except in the neighborhood of $a = 0$). There the PLKF errors are monotone decreasing over the incoming segment ($a > 0$), and they are monotone increasing over the outgoing segment ($a < 0$). Note also that this example the *rmse* at $a = 0$ has a "spike: such is caused by the rotation of the covariance matrix as the object moves past the radar.

In this example the principal axes of the measurements' covariance matrix rotate ninety degrees as the object approaches the radar, and they rotate another ninety degrees as the object moves away. Most of that rotation occurs from +60 to –60 degrees azimuth. Only four detection opportunities between those limits (overall, there are 156 detection opportunities). Apparently, the efficacy of the PLKF "debiasing" methods wane when the covariance matrix of the measurements is not constant (the non-stationary case).

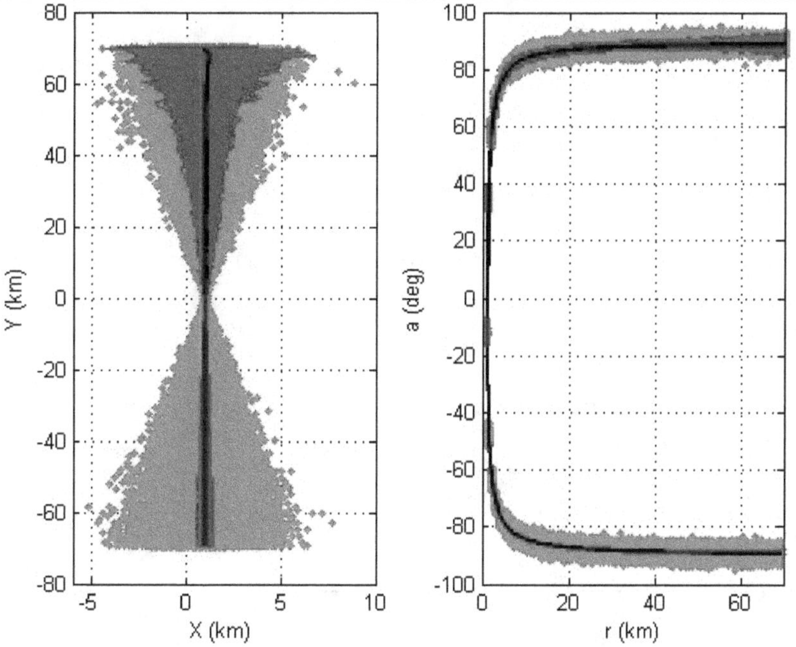

Figure 8-1 EPOT estimates for air and missile defense scenario ($\sigma_A = 1.5$)

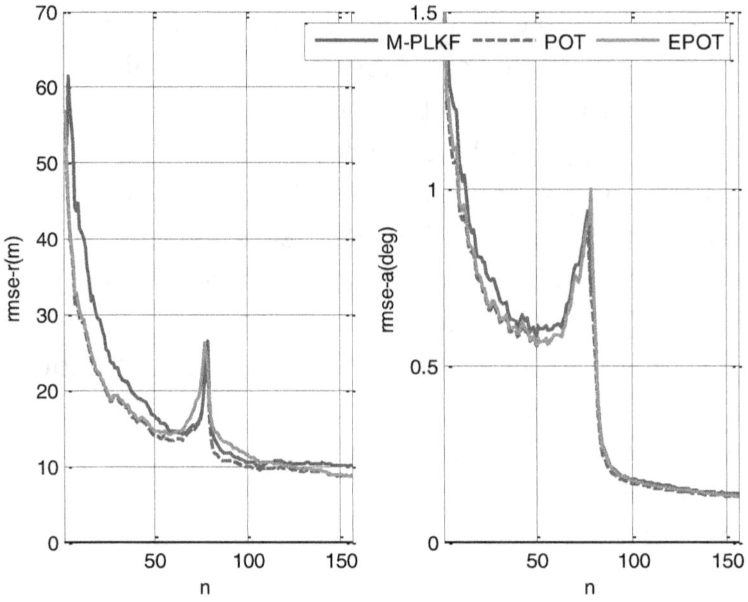

Figure 8-2 M-PLKF, POT, and EPOT root-mean-squared errors ($\sigma_A = 1.5$)

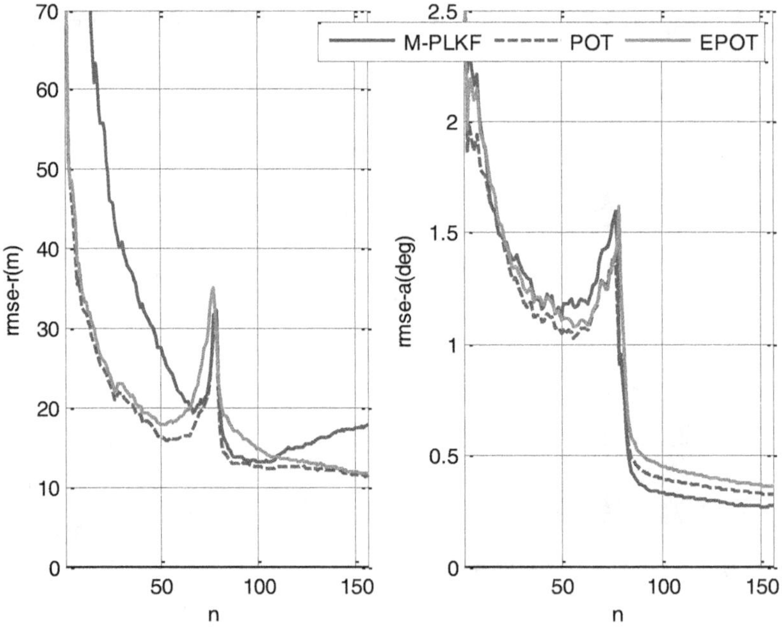

Figure 8-3 M-PLKF, POT, and EPOT root-mean-squared errors ($\sigma_A = 2.5$)

8.2.2 Maximum Likelihood Evaluation of the PLKF and EKF

The LKF is said to be optimal in the unbiased and minimum variance sense. However, its behavior in the *maximum likelihood* sense should also be considered. For example, let the object be motionless: fixed at $r = 70$ kilometers and $a = 0$ degrees, with $\dot{r} = \dot{a} = 0$; and then estimate its (constant) position and (zero) velocity – this is the CV case with zero velocity. Figure 8-4 provides the sets of tracks that are respectively determined by the PLKF, EKF, and the EKF with the POT and EPOT. The respective sets of tracks are overlaid upon the given measurements (the top plots are in $\mathbb{E}$ (Euclidean space) and the lower plots are in $\mathbb{R}$ (radar coordinate space).

The debiased and consistent" PLKF tracks are more likely to be biased toward the radar; the EKF tracks (without the POT or EPOT) are more likely to be biased away from the radar; and the tracks determined under the POT and EPOT are less likely to be biased.

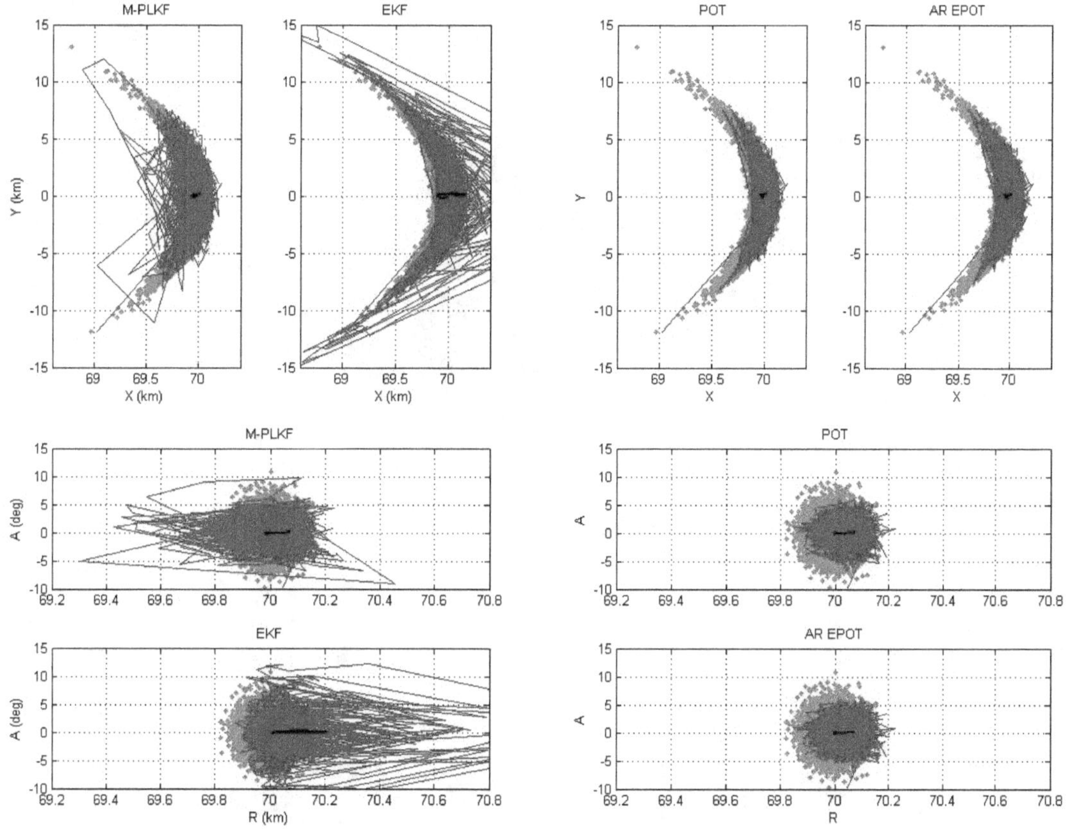

Figure 8-4 Sample distributions of measurements and estimates

9

Appendix

In this Appendix the equations for the *Best Linear Unbiased Estimator* (BLUE) that are used in this work are derived, along with its forms – the *batch* form, the recursive *fusion* and *update* forms. This presentation is based on our work in [7].

By definition, a *linear estimate* of $\mathbf{y}$ given $\bar{\mathbf{z}}$ has the form $\hat{\mathbf{y}} = \mathbf{A}\bar{\mathbf{z}}$, where $\mathbf{A}$ is a linear transformation that is independent of $\bar{\mathbf{z}}$. Here $\mathbf{y} \in \boldsymbol{Y}$ is some unknown to be determined. The corresponding affine form is $\bar{\mathbf{z}} = \mathbf{H}\mathbf{y} + \tilde{\mathbf{z}}$, with $\mathbf{z} = \mathbf{H}\mathbf{y}$ a linear transformation and $\tilde{\mathbf{z}}$ the measurement error vector. By assumption, $\tilde{\mathbf{z}}$ is independent of $\mathbf{z}$ and $\mathbf{H}$ has maximal rank such that $(\mathbf{H}^T\mathbf{H})^{-1}$ exists. In which case, the pseudo-inverse $\mathbf{A} = (\mathbf{H}^T\mathbf{H})^{-1}\mathbf{H}^T$ exists, dubbed the *linear least-squares* estimation matrix. The linear least-squares estimate of $\mathbf{y}$ given $\bar{\mathbf{z}}$ is then $\hat{\mathbf{y}} = (\mathbf{H}^T\mathbf{H})^{-1}\mathbf{H}^T\bar{\mathbf{z}}$.

Now to discuss the desired properties of such an estimate, that it be *unbiased* and have *minimum variance*, we invoke an underlying random vector for $\bar{\mathbf{z}}$, denoted $\boldsymbol{Z}$. The measurement vector $\bar{\mathbf{z}}$ is a realization of $\boldsymbol{Z}$. This in turn implies the underlying random vector of the measurement errors is $\tilde{\boldsymbol{Z}} = \boldsymbol{Z} - \mathbf{H}\mathbf{y}$, and we further assume that $\tilde{\boldsymbol{Z}}$ does not depend on $\mathbf{H}\mathbf{y}$. The least-squares linear transformation matrix $(\mathbf{H}^T\mathbf{H})^{-1}\mathbf{H}^T$ implies the underlying random vector of $\hat{\mathbf{y}}$ is $\mathcal{E}\boldsymbol{Y}$, where $\mathcal{E}$ denotes the (linear) *expectation operator*. The measurements are then unbiased if $\mathcal{E}\tilde{\boldsymbol{Z}} = \mathbf{0}$, equivalently $\mathcal{E}\boldsymbol{Z} = \mathbf{H}\mathbf{y}$, which also implies $\mathcal{E}\boldsymbol{Y} - \mathbf{y} = \mathbf{0}$. And the estimate $\hat{\mathbf{y}} = (\mathbf{H}^T\mathbf{H})^{-1}\mathbf{H}^T\bar{\mathbf{z}}$ is unbiased too. The estimate $\hat{\mathbf{y}} = \mathbf{A}\bar{\mathbf{z}}$ is a realization of $\boldsymbol{Y} = \mathbf{A}\boldsymbol{Z}$. And, by definition, the covariance matrix of $\tilde{\boldsymbol{Z}}$ is

$\operatorname{cov}\tilde{\boldsymbol{Z}} = \mathcal{E}\left(\tilde{\boldsymbol{Z}} - \mathcal{E}\tilde{\boldsymbol{Z}}\right)\left(\tilde{\boldsymbol{Z}} - \mathcal{E}\tilde{\boldsymbol{Z}}\right)^T$, when it exists. So $\operatorname{cov}\tilde{\boldsymbol{Z}} = \mathcal{E}(\tilde{\boldsymbol{Z}}\tilde{\boldsymbol{Z}}^T)$ since $\boldsymbol{Z}$ is unbiased. Our objective is for $\boldsymbol{Y} = \mathbf{A}\boldsymbol{Z}$ to have minimum variance, assuming that the covariance matrix of $\boldsymbol{Z}$ exists and is nonsingular (*positive definite*).

Now let a sequence of such $\boldsymbol{Z}$'s be given, $\boldsymbol{Z}_n = \mathbf{H}_n\mathbf{y} + \tilde{\boldsymbol{Z}}_n$, $n = 1, 2, \cdots, N$, with all the $\tilde{\boldsymbol{Z}}_n$ unbiased and mutually independent. This last requirement stipulates $\operatorname{cov}(\tilde{\boldsymbol{Z}}_m\tilde{\boldsymbol{Z}}_n^T) = \mathbf{0}$, the zero matrix, when $m \neq n$. The form of the *linear* least-squares estimator then generalizes to

$$\boldsymbol{Y}_N = \mathbf{A}_1\boldsymbol{Z}_1 + \mathbf{A}_2\boldsymbol{Z}_2 + \cdots + \mathbf{A}_N\boldsymbol{Z}_N, \tag{308}$$

a linear combination of the underlying random vectors. And for this $\boldsymbol{Y}_N$ to be *unbiased* we require

$$\mathcal{E}\boldsymbol{Y}_N = \sum_{n=1}^{N}\mathbf{A}_n\mathcal{E}\boldsymbol{Z}_n = \sum_{n=1}^{N}\mathbf{A}_n\mathbf{H}_n\mathbf{y} = \mathbf{y}. \tag{309}$$

Equivalently,

$$\sum_{n=1}^{N}\mathbf{A}_n\mathbf{H}_n = \mathbf{I}. \tag{310}$$

By definition, the covariance matrix of this estimate is

$$\operatorname{cov}\boldsymbol{Y}_N = \mathcal{E}\left(\sum_{m=1}^{N}\mathbf{A}_m\boldsymbol{Z}_m - \mathcal{E}\sum_{m=1}^{N}\mathbf{A}_m\boldsymbol{Z}_m\right)\left(\sum_{n=1}^{N}\mathbf{A}_n\boldsymbol{Z}_n - \mathcal{E}\sum_{m=1}^{N}\mathbf{A}_m\boldsymbol{Z}_m\right)^T. \tag{311}$$

And for $\boldsymbol{Y}_N$ to have *minimum variance*, the $\mathbf{A}_n$'s must be chosen such that the trace of $\operatorname{cov}\boldsymbol{Y}_N$ is minimized (assuming, of course, that $\operatorname{cov}\boldsymbol{Y}_N$ is positive definite, as well as all the $\operatorname{cov}\boldsymbol{Z}_n$). This is a constrained optimization problem: minimize the trace of (312) subject to the constraint in (310).

Now since the $\boldsymbol{Z}_n$ area all unbiased and mutually independent, the $\tilde{\boldsymbol{Z}}_n$ are all mutually orthogonal, In particular, $\mathcal{E}(\boldsymbol{Z}_m\boldsymbol{Z}_n^T) = \Sigma_Z(n)$ if $m = n$, else $\mathcal{E}(\boldsymbol{Z}_m\boldsymbol{Z}_n^T) = \mathbf{0}$. And so

$$\boldsymbol{\Sigma}_Y(N) = \sum_{n=1}^{N}\mathbf{A}_n\boldsymbol{\Sigma}_Z(n)\mathbf{A}_n^T. \tag{312}$$

Let $\boldsymbol{\Lambda}$ be a non-singular Lagrange multiplier matrix and consider

$$\lambda = \mathrm{tr}\left(\sum_{n=1}^{N} \mathbf{A}_n \boldsymbol{\Sigma}_{\boldsymbol{Z}}(n) \mathbf{A}_n^T\right) - 2\mathrm{tr}\left(\boldsymbol{\Lambda}^T \sum_{n=1}^{N} \mathbf{A}_n \mathbf{H}_n\right). \tag{313}$$

For each $n = 1, 2, \cdots, N$, the derivative of this scalar λ with respect to the matrix $\mathbf{A}_n$ is

$$\frac{\partial \lambda}{\partial \mathbf{A}_n} = 2\mathbf{A}_n \boldsymbol{\Sigma}_{\boldsymbol{Z}}(n) - 2\boldsymbol{\Lambda}\mathbf{H}_n^T.$$

Setting these expressions equal to the corresponding zero matrices,

$$\mathbf{A}_n = \boldsymbol{\Lambda}\mathbf{H}_n^T \boldsymbol{\Sigma}_{\boldsymbol{Z}}^{-1}(n). \tag{314}$$

And substituting the results back into (312), and noting that

$$\mathbf{A}_n \boldsymbol{\Sigma}_{\boldsymbol{Z}}(n) \mathbf{A}_n^T = \boldsymbol{\Lambda}\mathbf{H}_n^T \mathbf{A}_n^T = \mathbf{H}_n \mathbf{A}_n \boldsymbol{\Lambda}^T \tag{315}$$

we obtain

$$\boldsymbol{\Sigma}_{\boldsymbol{Y}}(N) = \boldsymbol{\Lambda} \sum_{n=1}^{N} \mathbf{H}_n^T \mathbf{A}_n^T = \sum_{n=1}^{N} \mathbf{A}_n \mathbf{H}_n \boldsymbol{\Lambda}^T. \tag{316}$$

But using (310) in (316) implies

$$\boldsymbol{\Sigma}_{\hat{\boldsymbol{Y}}}(N) = \boldsymbol{\Lambda} = \boldsymbol{\Lambda}^T. \tag{317}$$

So post-multiplying both sides of (314) by $\mathbf{H}_n$ and then summing yields

$$\boldsymbol{\Lambda} \sum_{n=1}^{N} \mathbf{H}_n^T \boldsymbol{\Sigma}_{\boldsymbol{Z}}^{-1}(n) \mathbf{H}_n = \sum_{n=1}^{N} \mathbf{A}_n \mathbf{H}_n = \mathbf{I}.$$

Thus,

$$\boldsymbol{\Sigma}_{\boldsymbol{Y}}(N) = \left(\sum_{n=1}^{N} \mathbf{H}_n^T \boldsymbol{\Sigma}_{\boldsymbol{Z}}^{-1}(n) \mathbf{H}_n\right)^{-1} = \boldsymbol{\Lambda}. \tag{318}$$

The $\mathbf{A}_n$'s that minimize the trace of (312) subject to (310) are thus

$$\mathbf{A}_n = \left(\sum_{n=1}^{N} \mathbf{H}_n^T \boldsymbol{\Sigma}_Z^{-1}(n) \mathbf{H}_n \right)^{-1} \mathbf{H}_n^T \boldsymbol{\Sigma}_Z^{-1}(n) = \boldsymbol{\Sigma}_{\hat{Y}}(N) \mathbf{H}_n^T \boldsymbol{\Sigma}_Z^{-1}(n) . \tag{319}$$

Therefore, the BLUE is

$$\hat{Y}_N = \sum_{n=1}^{N} \mathbf{A}_n Z_n = \left(\sum_{n=1}^{N} \mathbf{H}_n^T \boldsymbol{\Sigma}_Z^{-1}(n) \mathbf{H}_n \right)^{-1} \sum_{n=1}^{N} \mathbf{H}_n^T \boldsymbol{\Sigma}_Z^{-1}(n) Z_n . \tag{320}$$

The estimate and covariance matrix determined by the BLUE are

$$\hat{\mathbf{y}}_N = \boldsymbol{\Sigma}_Y(N) \sum_{n=1}^{N} \mathbf{H}_n^T \boldsymbol{\Sigma}_Z^{-1}(n) \overline{\mathbf{z}}_n \quad \text{and} \quad \boldsymbol{\Sigma}_Y(N) = \left(\sum_{n=1}^{N} \mathbf{H}_n^T \boldsymbol{\Sigma}_Z^{-1}(n) \mathbf{H}_n \right)^{-1} . \tag{321}$$

This form of the BLUE in (321) is a "batch" estimator: all the measurements are being used at once, concurrently in a single batch to determine the estimate. However, the BLUE may also be determined recursively as follows.

From (321), determine the sequences

$$\boldsymbol{\Sigma}_Y^{-1}(n) \hat{\mathbf{y}}_n \equiv \sum_{k=1}^{n} \mathbf{H}_k^T \boldsymbol{\Sigma}_Z^{-1}(k) \overline{\mathbf{z}}_k \quad \text{and} \quad \boldsymbol{\Sigma}_Y^{-1}(n) \equiv \sum_{k=1}^{n} \mathbf{H}_k^T \boldsymbol{\Sigma}_Z^{-1}(k) \mathbf{H}_k , \tag{322}$$

$n = 1, 2, \cdots, N$. For convenience, let $\boldsymbol{\Sigma}_Y^{-1}(0) \equiv \mathbf{0}$, and use the second expression in (322) to define $\boldsymbol{\Sigma}_{\hat{Y}}^{-1}(n)$. Whereupon, for $n = 1, 2, \cdots, N$ in (322), the BLUE can be determined recursively as

$$\boldsymbol{\Sigma}_Y^{-1}(n) \hat{\mathbf{y}}_n = \boldsymbol{\Sigma}_Y^{-1}(n-1) \hat{\mathbf{y}}_{n-1} + \mathbf{H}_n^T \boldsymbol{\Sigma}_Z^{-1}(n) \overline{\mathbf{z}}_n \tag{323}$$

and

$$\boldsymbol{\Sigma}_Y^{-1}(n) = \boldsymbol{\Sigma}_Y^{-1}(n-1) + \mathbf{H}_n^T \boldsymbol{\Sigma}_Z^{-1}(n) \mathbf{H}_n . \tag{324}$$

Note that for the batch BLUE estimate to exist, $\boldsymbol{\Sigma}_Y(N)$ must be nonsingular. Thus, there is a smallest $n \leq N$, say m, for which $\boldsymbol{\Sigma}_Y(m)$ is nonsingular. Accordingly, invert (324) at that m and then multiply $\boldsymbol{\Sigma}_Y^{-1}(m) \hat{\mathbf{y}}_m$ by $\boldsymbol{\Sigma}_Y(m)$. In which case

$$\hat{\mathbf{y}}_m = \boldsymbol{\Sigma}_Y(m) \left[\boldsymbol{\Sigma}_{\hat{Y}}^{-1}(m-1) \hat{\mathbf{y}}_{m-1} + \mathbf{H}_m^T \boldsymbol{\Sigma}_Z^{-1}(m) \overline{\mathbf{z}}_m \right] . \tag{325}$$

And for $n > m$ substitute (324) as $\mathbf{\Sigma}_Y^{-1}(n-1) = \mathbf{H}_n^T \mathbf{\Sigma}_Z^{-1}(n)\mathbf{H}_n - \mathbf{\Sigma}_{\hat{Y}}^{-1}(n)$ into (325). Whereupon, the recursive form of the BLUE is obtained to be

$$\hat{\mathbf{y}}_n = \hat{\mathbf{y}}_{n-1} + \mathbf{\Sigma}_Y(n)\mathbf{H}_n^T \mathbf{\Sigma}_Z^{-1}(n)\left(\overline{\mathbf{z}}_n - \mathbf{H}_n \hat{\mathbf{y}}_{n-1}\right) \tag{326}$$

and

$$\mathbf{\Sigma}_Y(n) = \left[\mathbf{\Sigma}_Y^{-1}(n-1) + \mathbf{H}_n^T \mathbf{\Sigma}_Z^{-1}(n)\mathbf{H}_n\right]^{-1}. \tag{327}$$

The more commonly used form of the recursive BLUE is

$$\hat{\mathbf{y}}_n = \hat{\mathbf{y}}_{n-1} + \mathbf{K}_n\left(\overline{\mathbf{z}}_n - \mathbf{H}_n \hat{\mathbf{y}}_{n-1}\right) \text{ and } \mathbf{\Sigma}_Y(n) = \left(\mathbf{I} - \mathbf{K}_n \mathbf{H}_n\right)\mathbf{\Sigma}_Y(n-1) \tag{328}$$

where

$$\mathbf{K}_n = \mathbf{\Sigma}_Y(n-1)\mathbf{H}_n^T\left[\mathbf{H}_n \mathbf{\Sigma}_Y(n-1)\mathbf{H}_n^T + \mathbf{\Sigma}_Z(n)\right]^{-1}, \tag{329}$$

called the gain matrix– $\mathbf{\Sigma}_Y(n)\mathbf{H}_n^T \mathbf{\Sigma}_Z^{-1}(n) = \mathbf{K}_n$ in (324). These last to forms of the BLUE are related by the so-called matrix inversion lemma [2]: the recursive form given by (326) and (327) is more amenable to analysis; the recursive form given by (328) and (329) is usually less computationally burdensome.

9.1 Appendix References

[1] K. S. Miller and D. M. Leskiw, An Introduction to Kalman Filtering with Applications, Krieger (1987).

[2] K. S. Miller, Some Eclectic Matrix Theory, Krieger (1987).

www.ingramcontent.com/pod-product-compliance
Lightning Source LLC
Chambersburg PA
CBHW081001170526
45158CB00010B/2865
* 9 7 8 1 0 9 2 9 5 4 5 1 8 *